STATICS FOR BUILDING CONSTRUCTION

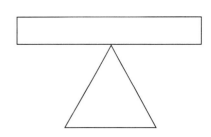

STATICS FOR BUILDING CONSTRUCTION

Jack L. Burton

Prentice Hall
Upper Saddle River, New Jersey   Columbus, Ohio

Library of Congress Cataloging-in-Publication Data

Burton, Jack L.
    Statics for building construction / Jack L. Burton.
        p. cm.
    Includes bibliographical references.
    ISBN 0-13-674300-5
      1. Structural analysis (Engineering)   2. Statics.   I. Title.
TA648.B87   1998
624.1'71—dc21   98-9661
                                                                                                     CIP

Editor: Ed Francis
Production Coordinator: Christine M. Harrington
Production Editor: Karen Fortgang, bookworks
Design Coordinator: Karrie M. Converse
Text Designer: STELLARViSIONs
Cover Designer: Raymond Hummons
Production Manager: Patricia A. Tonneman
Marketing Manager: Danny Hoyt

This book was set in New Century Schoolbook by STELLARViSIONs and was printed and bound by Courier/Kendallville, Inc. The cover was printed by Phoenix Color Corp.

© 1999 by Prentice-Hall, Inc.
Simon & Schuster/A Viacom Company
Upper Saddle River, New Jersey 07458

All rights reserved. No part of this book may be reproduced, in any form or by any means, without permission in writing from the publisher.

Printed in the United States of America

10 9 8 7 6 5 4 3 2 1

ISBN 0-13-674300-5

Prentice-Hall International (UK) Limited, *London*
Prentice-Hall of Australia Pty. Limited, *Sydney*
Prentice-Hall of Canada, Inc., *Toronto*
Prentice-Hall Hispanoamericana, S. A., *Mexico*
Prentice-Hall of India Private Limited, *New Delhi*
Prentice-Hall of Japan, Inc., *Tokyo*
Simon & Schuster Asia Pte. Ltd., *Singapore*
Editora Prentice-Hall do Brasil, Ltda., *Rio de Janeiro*

*To my wife, Tracy,
my parents, Melvin and Helen,
and my Lord and Savior, Christ Jesus,
without all of whom this book would not have been possible.*

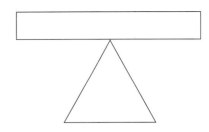

Preface

This textbook is intended to be an instructional text for **Strength of Materials** courses offered at the undergraduate college level and for similar courses offered in trade and technical schools. The student using this text in conjunction with such a course should first have a basic understanding of algebraic manipulation and trigonometric functions. Any technical mathematics courses containing these elements will suffice for this purpose. In lieu of the availability of these courses, it is recommended that the student take Algebra I and Basics of Trigonometry, or similar courses.

Many possible approaches can be used to study the topics introduced in this text. Because of my familiarity with architectural structures, I have chosen to use architectural structural analysis. The course has five major areas of study: static physics, analysis of forces on beams, structural steel design, structural wood design, and structural concrete design.

All computations covered are based on two-dimensional forces in order to simplify the scope of this study. Problems and examples have been provided for the use of instructors and students alike. Instructors are encouraged to provide real-life examples as they apply. Additionally, short- and long-term projects have been provided at the back of this text. The order of the classroom projects matches the order in which those principles appear in this text.

The purpose of this text is to guide the student into the analytical thought process required to make an informed structural analysis. The instructor is encouraged to teach this course in such a manner. Classroom props and class projects may be used to aid the student in the understanding of the concepts taught in this course. Most of the scenarios portrayed in this text are in the form of word problems. This is done to acclimate the student to real-life applications of the material discussed.

It is the sincere desire of this author that the material presented in this text be useful to both student and instructor alike. Every effort has been made to clarify all concepts presented herein. All ANSI (American National

Standards Institute) and ASTM (American Society for Testing and Materials) standards have been adhered to wherever applicable. Standard mathematical convention is used throughout this text.

Jack L. Burton

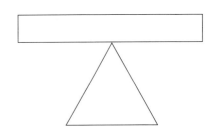

Acknowledgments

Assistants to the Author
Brenda Barto
Chantal Sparrow

Graphic Artists
Jonathan Fritz
Brian Hoerner
Troy Shirey
George Windemaker, Sr.

3D Graphic Artist
Brian Hoerner

Photographer
Brenda Barto

Reviewers
Ron Gallagher, University of Toledo
John Jarchow, Pima Community College

Proofreaders
Wendy Arter
Tracy A. Burton
Aaron Horton
Sabrina Johnson
Sandra McNeal
Nelson Sierra
Chantal Sparrow
Tony Thomas

Technical Support
Brenda Barto
Slade Bush
Jonathan Fritz
Sabrina Johnson
Sandra McNeal
Chantal Sparrow

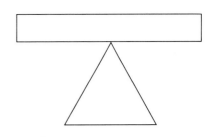

Contents

Preface   vii

Acknowledgments   ix

Introduction   1

Chapter 1   Forces: Acting and Reacting   5

1.1   Introduction   5
1.2   Forces Denied   5
1.3   Reacting Forces   7
1.4   Properties of Forces   9
1.5   Summary   10

Chapter 2   Analysis of Vectors in Two Dimensions   17

2.1   Introduction   17
2.2   Vectors Defined   17
2.3   Vector Components   18
2.4   Resultant Form of Vectors   22
2.5   Summary   24

Chapter 3   Work and Energy   39

3.1   Introduction   39
3.2   Work Defined   39
3.3   Additional Work Scenarios   40
3.4   Potential Energy   46
3.5   Kinetic Energy   49
3.6   Summary   53

Chapter 4 Moments 61

- 4.1 Introduction 61
- 4.2 Moments Defined 61
- 4.3 Convention of Moments 62
- 4.4 First-Order Levers 64
- 4.5 Centroids 67
- 4.6 Second-Order Levers 70
- 4.7 Stabilizing Forces 72
- 4.8 Normal Forces 76
- 4.9 Angled Stabilizing Forces 79
- 4.10 Summary 82

Chapter 5 Analysis of Beam Forces and Reactions 93

- 5.1 Introduction 93
- 5.2 Beam Forces and Reactions Defined 93
- 5.3 Computation of Beam Reactions 94
- 5.4 Computation of Column Compressions due to Internal Forces 101
- 5.5 Computation of Column Compressions due to External Forces 103
- 5.6 Summary 107

Chapter 6 Steel Beam Design 119

- 6.1 Introduction 119
- 6.2 Steel Terminology 119
- 6.3 Properties of Metals 121
- 6.4 Architectural Steel: Properties and Stress Analysis 122
- 6.5 Uniform Load Distributions on Rectangular Sections 127
- 6.6 Steel Beam Nomenclature 132
- 6.7 Cutting Lengths of Steel Beams 135
- 6.8 Sizing Steel I-beams Using Standard Live-Load Deflection 139
- 6.9 Computation of Maximum Allowable Live Loads on Beams 141
- 6.10 Summary 142

Chapter 7 Steel Joist and Fastener Design 151

- 7.1 Introduction 151
- 7.2 Steel Framing Plans 152
- 7.3 Sizing Joists 153
- 7.4 Clip Selection 155
- 7.5 Bolt/Rivet Shear 157
- 7.6 Computation of Load Safety Factors 159
- 7.7 Clip Detailing 161
- 7.8 Custom Beam Cuts 164
- 7.9 Assembling the Steel Beam Detail 169
- 7.10 Summary 170

Chapter 8 Wood Joist and Column Design and Detailing   179

8.1 Introduction   179
8.2 Computation of Moment of Inertia   179
8.3 Wood Nomenclature   184
8.4 Sizing Single Wood Members   187
8.5 Sizing Multiple Wood Members   192
8.6 Deck and Floor Section Design   195
8.7 Solid Sawn Wood Columns   198
8.8 Summary   202

Chapter 9 Reinforced Concrete Columns and Footings   209

9.1 Introduction   209
9.2 Rebar Nomenclature and Strength   209
9.3 Column Eccentricity   212
9.4 Sizing Square and Round Reinforced Concrete Columns   215
9.5 Sizing for Compressions on Concrete Footings   219
9.6 Detailing Reinforced Concrete Columns and Footings   223
9.7 Summary   226

Class Projects   231

Tables   251

Glossary   305

Bibliography   313

Index   315

STATICS FOR BUILDING CONSTRUCTION

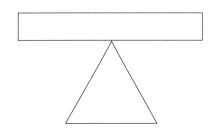

Introduction

Several basic concepts concerning the study of **strength of materials** (sometimes referred to as *mechanics of materials*) need to be established before the main body of this course can be properly approached. In this section we discuss the core topics around which this course has been designed.

The first concept to be established is that of **static equilibrium**. The word *static* means stationary or unmoving, and *equilibrium* refers to a state of balance. A system is in static equilibrium when there is no motion within the system as a result of internal and/or external forces. For a system to be in static equilibrium the following criteria must be true:

1. The sum of all the forces internal and external to the system must be zero.
2. The sum of all the work done on and by the system must be zero.
3. The sum of all the moments acting within the system must be zero.
4. The sum of all the reactions within the system must be equal to the sum of all the actions upon the system.

As we move through this course, all of the above criteria will be established mathematically and practically.

The main purpose of this course is to teach the student to investigate the **structural integrity** of a system and to complete a proper **structural analysis** of that system. Study the definitions of each of these terms given below:

Structural integrity—the ability of an object or system to resist forces internally and externally applied on that system (the ability of a system to do what it has been designed to do).

Structural analysis—the study of the materials, type and purpose of structure, and the acting and reacting forces within and upon a system. Structural analysis requires that three criteria be known:

1. Kind(s) of force(s) acting on the system
2. Intensity (magnitude) of force(s) acting on the system
3. Sense (direction) of force(s) acting on the system

A proper structural analysis of any system is guided by, but is not necessarily limited to, these criteria.

All problems and information presented in this text follow the pattern given below:

1. **Theory:** Basic concepts and definitions will be addressed during the introduction of each topic.
2. **Principle:** Formulas and mathematical relationships will be established for each topic discussed.
3. **Application:** Real-life scenarios will be explored wherever possible during the study of each topic in this text.

Several mathematical symbols and abbreviations are used in the study of strength of materials. A comprehensive list of these symbols along with their meanings is given below:

a	arbitrary fixed point on a moment arm, fulcrum
B	cut steel beam length
b_f	steel beam flange width
bf	board-feet
D	beam deflection (ratio)
d	distance, steel beam depth
E	modulus of elasticity
$\mathbf{F}$	force
F_x	horizontal component of a force
F_y	vertical component of a force
$\mathbf{F}_T$	vector sum of forces
g	gravitational acceleration constant
I	moment of inertia
I_x	moment of inertia about x axis
I_y	moment of inertia about y axis
L	distance between column center lines
l	cut steel beam length in deflection formula
M	moment
M_a	moment about point a
psi	pounds per square inch
R	resultant

$R_a \Rightarrow M_b$	beam reaction at point a due to the moment about point b
sf	square feet (foot)
s.f.	safety factor
t_w	steel beam web thickness
v	velocity
W	work
α	acute angle with x axis (alpha—pronounced ALFA)
δ	beam deflection (deviation) in inches (lower case delta)
ψ	acute angle with a defined surface (psi—pronounced SIGH)
Δ	"the change in ..." (upper case delta)
θ	angle with positive x axis (theta—pronounced THAY-TUH)
π	3.1415926 ... (pi—pronounced PIE)
Σ	"the sum of ..." (sigma)
±	plus or minus
≤	less than or equal to
≥	greater than or equal to
|a|	absolute value of a
≈	approximately equal to
∥	parallel
⇒	leading to, due to
∋	such that
≠	not equal to
∴	therefore
∡	angle symbol
$\stackrel{?}{=}$	questioned equality
∟	right angle
⊥	perpendicular to

It is recommended that the student become familiar with each of these symbols and abbreviations, since they are used throughout this text.

CHAPTER 1

Forces: Acting and Reacting

1.1 INTRODUCTION

In this chapter, we examine the following topics as they are related to forces in a two-dimensional system:

1. The definition of a force
2. Four kinds of acting and reacting forces
3. How forces interact within the same plane
4. Internal and external forces
5. The role that acting and reacting forces play in a structural analysis

1.2 FORCES DEFINED

For the purpose of this course, we define a **force** as a motivator that acts along a specific line in a specific direction that tends to bring about a change in the state of a system. For example, if a soccer ball lies at rest on the ground, then it is in static equilibrium. Its current state remains unchanged until it is kicked by a player. The foot of the player provides the motivation, that is, the force, to propel the ball, which is then no longer in static equilibrium, since it is now moving. The direction in which the ball is propelled indicates the direction of the force imparted on it.

A force is made up of two parts, magnitude and direction:

1. **Magnitude**—the intensity of the force, usually measured in pounds, kilopounds (kips), tons (2000 lb), kilograms, or similar units
2. **Direction**—the angle commonly measured from the horizontal or from a chosen line at which a force is applied. This definition is more completely explained in the next chapter.

So, we say that every force has a magnitude and a direction. There is also one more thing we must know about a force, and that is the kind or type of force that it is. Kinds of forces are defined according to the way in which they act upon an object. Four kinds of **acting forces** are examined in this text:

1. **Tension:** a force that tends to cause elongation of an object along an axis
2. **Compression:** a force that tends to cause the shortening of an object along an axis
3. **Shear:** a force that acts at a sharp angle to an object and tends to cause that object to break across an axis
4. **Torsion** (also called **torque**): a force that causes an object to twist or rotate about an axis

---

A note about friction: Friction is often treated as a separate force in the study of physics, but, it is actually the result of thousands of microscopic shear forces occurring at the same time. Look at Figure 1.1. The block seems to "stick" to the surface when pushed, because, on the microscopic level, thousands of imperfections in the surface bump against one another. This causes a resistance which we call **friction**.

FIGURE 1.1

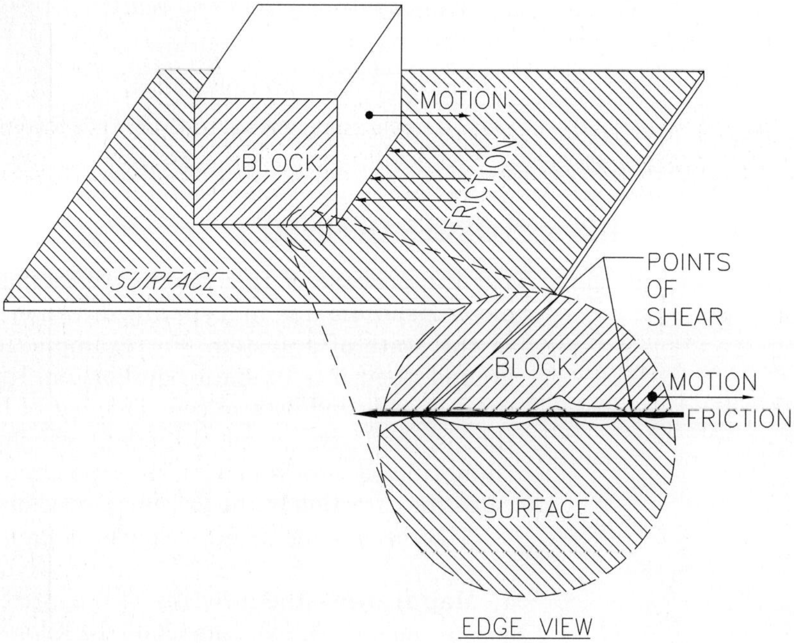

1.3 REACTING FORCES

For each of the acting forces there are also stabilizing forces known as **reacting forces**:

1. **Tensile strength:** the ability of an object to resist elongation along an axis
2. **Compressive strength:** the ability of an object to resist being shortened along an axis
3. **Breaking strength:** the ability of an object to resist breakage across an axis
4. **Torsional strength:** the ability of an object to resist being twisted about an axis

The four resistive forces are examined to determine the structural integrity of an object or a system.

Bridges such as this one experience a variety of internal and external forces. The wall-like support resting in the river is mainly being subjected to compression from the bridge above. Notice that the support has been oriented so that its longest dimension runs parallel to the flow of the river. This has been done to reduce the amount of shear on the wall-like support.

This photograph is a close-up view of a concrete beam resting atop two concrete columns that support a bridge. The beam is experiencing shear and compression and the columns are being subjected mostly to compression. Notice that the beams are thicker at the points where they meet the columns. Why do you think this has been done?

These window overhangs must be mounted against the wall so that they do not slide off. The fastening method must have adequate breaking strength to hold the overhangs in place since the weight of these structures causes shear at the wall.

The common door hinge is constantly being subjected to torsion. The obvious torque occurs when we open or close a door, but there is another condition that causes torque at the hinges of a door. What would happen if you took one or more of the hinges off this door? Would rotation occur? How?

The posts on this colonial-style structure have been designed to support the extended porch roof above. These columns are being compressed by the weight of the overhanging roof. Can you think of any other factors that might contribute to the compression on these posts?

1.4 PROPERTIES OF FORCES

Every force acts in a single direction, which is represented with a line. The line representing any single force is known as a **line of force** (LOF). Lines of force may interact within the same plane in a number of ways. We examine three of these ways in this section.

Colinear forces act along the same line of action. Colinear forces may interact with one another in one of two ways:

Additive colinear forces act along the same line in the same direction.

Negating colinear forces act along the same line in opposite directions.

Coplanar forces act within the same plane. For the purposes of this text, all of the forces we deal with will be coplanar. **Coincident** forces all act on the same point. Chapter 2 deals exclusively with these kinds of forces.

The sources from which forces may act upon a system may be either internal or external to the system:

Internal forces act within a system due to the weight of the members that make up that system.

External forces originate outside the system itself.

For example, if a 1000-lb block rests on a beam that weighs 120 lb, the 120-lb force on the columns would be due to internal forces, while the 1000-lb force on the columns would be due to an external force. (See Figure 1.2.) External forces can be physically removed from a system without changing the structural integrity of that system. Internal forces cannot be removed from a system without making a change in that system.

Acting forces tend to move the system out of static equilibrium. For example, a bowling ball at rest is in static equilibrium until the bowler picks up the ball and rolls it down the alley. The weight of the ball and the friction between the ball and the alley are all reacting forces. In this case, however,

FIGURE 1.2

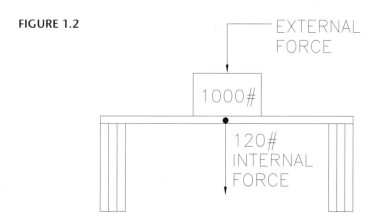

the reacting forces are not adequate to negate the acting forces exerted by the bowler. On the other hand, if someone were to cement the ball to the ball return, this might provide an adequate reacting force against the bowler's attempt to lift the ball.

When considering acting forces on a system, we must visualize the entire system, usually using a picture or sketch. We must determine what is taking place at all points of contact as well as at all weak spots within the system. Let us do a brief preliminary structural analysis on a seesaw:

ACTING FORCES
1. Weight of the seesaw board
2. Weight of the children on the board

REACTING FORCES
1. Compressive strength of the center support
2. Breaking strength of the board

Certain factors were not considered, such as

> Strength of the hinge
> Strength of the ground fasteners
> Speed and force with which the children play
> Atmospheric conditions

1.5 SUMMARY

We need to be aware of the **scope** of the analysis that we are about to perform. To correctly identify the roles of the acting and reacting forces within a system, we must first correctly identify the limits of our study based on the available and requested information. In the seesaw analysis we are interested only in determining the strength of the board and the support under a limited set of circumstances.

Study this section carefully. In the next chapter we examine structural systems more closely and use equations and principles rooted in physics to perform a more complete analysis of a system.

Use the analysis outline shown below when analyzing a system:

1. **Material(s)** List the materials used in the structure.
2. **Purpose** Define the purpose for which the structure has been designed.
3. **Forces**
    (a) *Kind* Identify the type(s) of force(s)
    (b) *Direction* Indicate the direction(s) of the force(s)
    (c) *Sense* Decide whether the force is moving toward or away from a given point

4. **Scale**

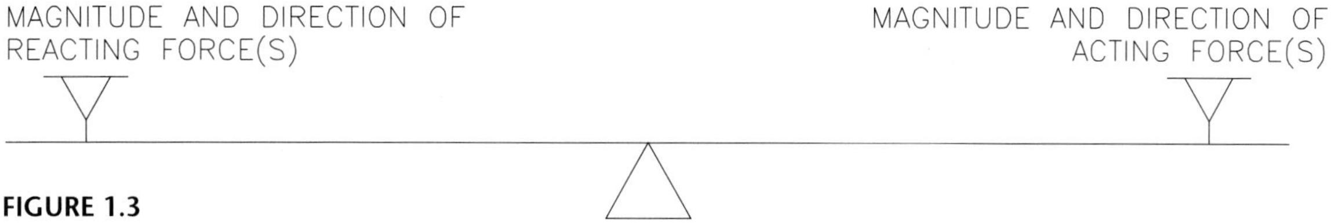

MAGNITUDE AND DIRECTION OF
REACTING FORCE(S)

MAGNITUDE AND DIRECTION OF
ACTING FORCE(S)

FIGURE 1.3

FIGURE 1.4

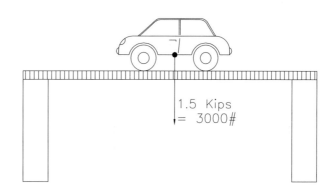

1.5 Kips
= 3000#

Sample Exercise 1.1

A 1.5-ton car is driven across a concrete bridge (Figure 1.4). Perform a structural analysis on this system. We use the outline format provided in the last section.

1. **Material(s)**   The bridge is made of concrete.
2. **Purpose**   To support the 3000-lb car.
3. **Forces**
   (a) *Kind*   The bridge surface is experiencing **shear**.
   (b) *Direction*   The acting force is being applied **downward** by the car
   (c) *Sense*   The force is acting **away** from the car.
4. **Scale**

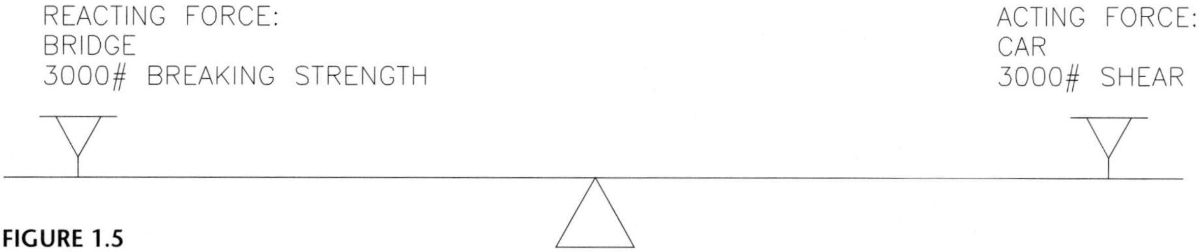

REACTING FORCE:
BRIDGE
3000# BREAKING STRENGTH

ACTING FORCE:
CAR
3000# SHEAR

FIGURE 1.5

FIGURE 1.6

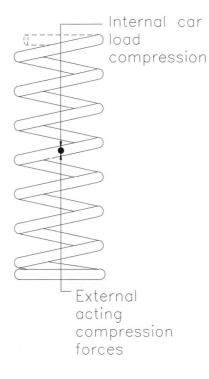

Automobile suspension spring.

Sample Exercise 1.2

An automobile suspension spring, shown in Figure 1.6 and in the photograph, must carry varying loads based on changing road conditions. Perform a structural analysis on this spring.

1. **Material(s)** — The spring is made of steel.
2. **Purpose** — To support the weight of the vehicle plus the varying loads caused by road conditions.
3. **Forces**
   (a) *Kind* — The spring experiences **compression**.
   (b) *Direction* — The acting forces are being applied **downward** by the car and **upward** by the road.
   (c) *Sense* — All acting forces are **toward** the center of the spring.
4. **Scale**

REACTING FORCE:
SPRING
COMPRESSIVE STRENGTH

ACTING FORCES:
CAR & ROAD CONDITIONS
COMPRESSION

FIGURE 1.7

FIGURE 1.8

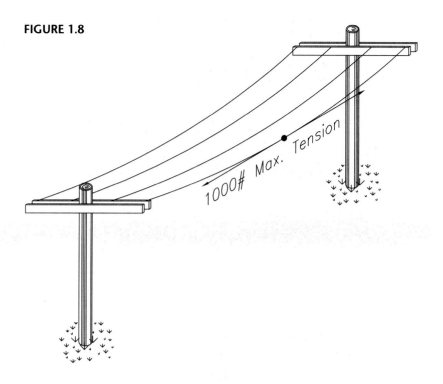

Sample Exercise 1.3

A 1000-lb aluminum power cable stretches between two poles as shown in Figure 1.8. Perform a structural analysis on the cable in this system.

1. **Material(s)**   The cable is made of aluminum.
2. **Purpose**   To support its own weight of 1000 lb.
3. **Forces**
    (a) *Kind*   The cable is under **tension**.
    (b) *Direction* The acting force is being applied **downward** by the weight of the pole causing it to elongate horizontally.
    (c) *Sense*   The force is acting **away** from the center of the cable.
4. **Scale**

REACTING FORCE:
CABLE
1000# TENSILE STRENGTH

ACTING FORCE:
CABLE
1000# TENSION

FIGURE 1.9

Problems

Using the analysis outline discussed in this chapter perform a structural analysis on the lettered element in each system in Problems 1–3.

1.

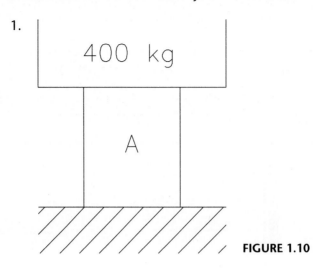

FIGURE 1.10

2.

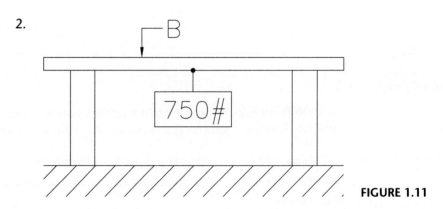

FIGURE 1.11

3.

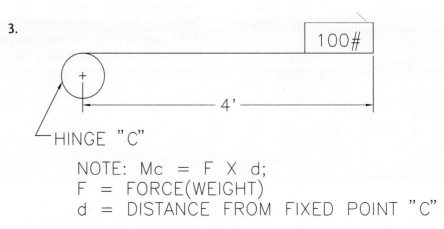

NOTE: Mc = F X d;
F = FORCE(WEIGHT)
d = DISTANCE FROM FIXED POINT "C"

FIGURE 1.12

4. List three examples in your everyday life in which you see or use each of the four forces discussed in this chapter.

5. Name the colinear forces that cause a car to accelerate on a road. Are these forces additive or negating?

6. When you sit in a chair what are the internal and external forces? Do these forces change as you get out of the chair? Why or why not?

7. What happens to the ground each time you take a step? Where do the reacting forces go when your foot leaves the ground?

Analysis of Vectors in Two Dimensions

2.1 INTRODUCTION

In this chapter we examine the following topics as they relate to forces in a single plane:

1. Vectors and vector sums
2. Vectors expressed in
   (a) Component form
   (b) Resultant form
3. Static equilibrium conditions for vectors
4. Practical applications for using vectors and vector sums

2.2 VECTORS DEFINED

A **vector** is a line drawn to represent a force. Since a vector is a graphic representation of a force, it also has two parts as defined in the previous chapter:

1. **Magnitude**: the numerical value of the intensity of the force
2. **Direction**: the angle measured counterclockwise from the positive x axis at which the force is acting

2.3 VECTOR COMPONENTS

In the last chapter we examined the effects that colinear forces have on one another. Of course, most forces do not act along a single line. In reality, forces act at odd angles to one another. These forces cannot be added in a simple manner. They must first be broken down into others forces, each of which act along the same line of action.

We call these parts of a vector its **components**. Every vector has two components, an x component and a y component. The components of a vector are represented by projecting a line from the end of a vector perpendicular to both the x axis and the y axis as shown in Figure 2.1. The components of a vector can be best understood if they are treated as the two legs of a right triangle. The x component is the horizontal leg and the y component is the vertical leg. The relationship between the vector and its components can be represented by the Pythagorean theorem: $\mathbf{F}_T^2 = F_x^2 + F_y^2$, where $\mathbf{F}_T$ is the vector represented by the hypotenuse of the triangle.

A **force triangle** can also be constructed to show the relationship between the sides and a given angle, as shown in Figure 2.2. Since F_x is the side adjacent to angle α then $\cos \alpha = F_x/F_T$. This leads to the relationship

$$F_x = F_T \cos \alpha$$

Subsequently, F_y is the side opposite angle α, so that $\sin \alpha = F_y/F_T$ and

$$F_y = F_T \sin \alpha$$

FIGURE 2.1

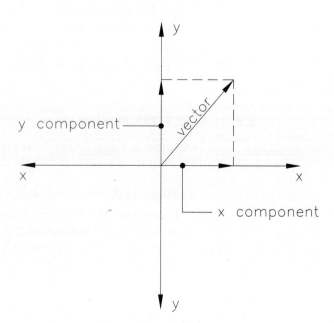

FIGURE 2.2

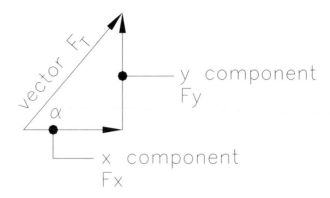

So, if we know the magnitude of a vector F_T and the angle α it makes with the horizontal, then we can find its components. A vector represented in this way is said to be expressed in component form.

Vectors do not always travel up and to the right, as shown in Figure 2.2; they may travel at any angle. In Figure 2.3, a representation of the Cartesian coordinate system is shown with quadrants and component properties labeled. As you can readily see, the quadrant in which a vector falls can be determined by the sign values of its components. For example, if the components of a particular vector are

$$F_x = -14.78 \text{ lb}$$
$$F_y = 29.23 \text{ lb}$$

then the vector would be located in the second quadrant, because the x component is negative in value and the y component is positive in value. Referring to Figure 2.3, we see that this occurs only in the second quadrant.

An example of angled forces in real life is demonstrated by the guy wires used to steady this power line pole. The guy wires are in tension as they hold the pole in place, but only the horizontal components of their tension prevent the pole from swaying. The amount of horizontal stabilizing force being applied by each guy wire is determined by (1) the wire's length, (2) the tensile strength of the wire, (3) the angle at which the wire is run. These factors can be used to determine where the guy wire should be attached to the pole and to the ground.

FIGURE 2.3

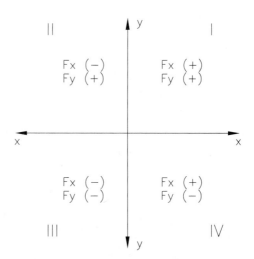

The direction of a vector will also indicate the quadrant in which it falls. As previously noted, the direction of a vector is measured counterclockwise from the positive side of the x axis. In this text we will refer to this angle as angle θ. Angle θ will always be positive in value, and will be given as an angle between 0 and 360°. The angle that the vector makes with the closest side of the x axis we will refer to as angle α, shown in Figure 2.2. Angle α will always be given as an acute angle and will, therefore, always have a value between 0 and 90°. The relationship between angle θ and angle α changes from quadrant to quadrant:

Quadrant	$\theta =$
I	α
II	$180 - \alpha$
III	$180 + \alpha$
IV	$360 - \alpha$

The purpose for breaking a given vector into components is to make it possible to add noncolinear vectors. Vectors that do not lie along the same line of action cannot be added directly. However, since all horizontal components lie along the same line, the x axis, these components can be added together just like any other colinear vectors. Likewise, vertical components can be added together as well.

Let us consider an example to investigate how this works. In solving a particular problem, you find two forces acting on a single point. The first force has a magnitude of 150 lb and acts at an angle of 30° from the horizontal in the first quadrant. The second force has a magnitude of 100 lb acting at an angle of 50° from the vertical in the second quadrant.

The first step in solving this type of problem is to draw a **pictorial diagram**, a sketch that represents the problem in picture form. Since the forces

FIGURE 2.4

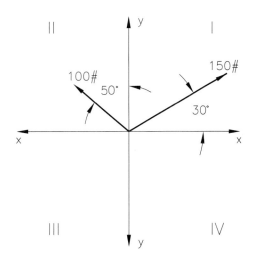

given in this example are relatively simple, they can be easily placed on a Cartesian grid as shown in Figure 2.4. The next step is to draw a **point-force diagram** on which all of the forces are labeled. In addition, the value for α is found for each force so that its components can be computed. In this example, one of the angles was given as an angle measured from the vertical. That angle must first be converted to an angle measured from the horizontal. Figure 2.5 shows the correct point-force diagram for this problem.

In the next step we compute F_x and F_y for each of the individual forces in this example. We then enter these values into a chart in order to find ΣF_x and ΣF_y. This gives us the vector sum expressed in **component form**. The components of force A are:

$$F_x = 150 \text{ lb} \times \cos 30° = 129.90 \text{ lb}$$
$$F_y = 150 \text{ lb} \times \sin 30° = 75.00 \text{ lb}$$

FIGURE 2.5

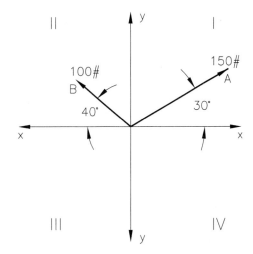

TABLE 2.1

Force	F_x (lb)	F_y (lb)
1	129.90	75.00
2	−76.60	64.28
	ΣF_x = 53.30 lb	139.28 lb = ΣF_y

The components of force B are:

$$F_x = -100 \text{ lb} \times \cos 40° = -76.60 \text{ lb}$$
$$F_y = 100 \text{ lb} \times \sin 40° = 64.28 \text{ lb}$$

Notice that the x component of force B is negative in value. That is because the x component of a vector in the second quadrant is negative (see Figure 2.3).

To find the sum of components we use a chart like Table 2.1. A vector sum expressed in the manner shown in this table is said to be expressed in component form. That is, the vector sum is given as the sums of its individual horizontal (x) and vertical (y) components.

2.4 RESULTANT FORM OF VECTORS

There is another, more convenient way to express a vector sum. We can determine the magnitude and direction of a vector sum if we know the sum of the components of two or more vectors.

Let us use as our example, the components of the problem shown on the previous page. If we draw the two components of this vector sum on a Cartesian coordinate system, we see that both vector sum components are positive in value, as shown in Figure 2.6. From this we conclude that the resultant vector that they represent will fall in the first quadrant (see Figure 2.3).

FIGURE 2.6

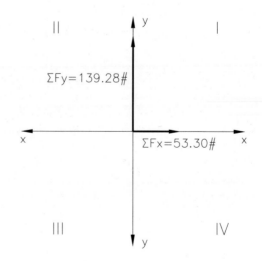

FIGURE 2.7

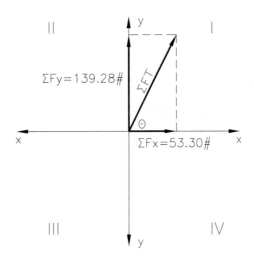

Next, we sketch in the resultant component by using the parallelogram method. This method is shown in Figure 2.7. The resulting vector represents the line of magnitude and direction of the vector sum. At this point, we need to find ΣF_T, which is the magnitude of the resulting force. We do this by using the Pythagorean theorem:

$$\Sigma F_T^2 = \Sigma F_x^2 + \Sigma F_y^2$$

Rewriting this formula and substituting the values found for ΣF_x and ΣF_y we get

$$\Sigma F_T = \sqrt{53.30^2 + 139.28^2}$$

Solving for ΣF_T we find $\Sigma F_T = 149.13$ lb. This is the magnitude of the vector sum.

Next, we need to find the angle θ that the vector sum makes when measured counterclockwise from the positive x axis. To do this we use the following formula:

$$\alpha = \tan^{-1} \left| \frac{\Sigma F_y}{\Sigma F_x} \right|$$

Once we have found the value for α, we use Table 2.1 to find the value for θ.

Substituting the known values for ΣF_x and ΣF_y into the formula above, we have

$$\alpha = \tan^{-1} \left| \frac{139.28}{53.30} \right|$$

$$= \tan^{-1} |2.61|$$

$$= 69.06°$$

The various forces caused by the weight of the object being lifted, the weight of the crane arm, and the stresses on the components of the crane all combine to create a resultant force that must be planned for by the designers of this piece of construction equipment. *Courtesy of MML Construction.*

Since $\theta = \alpha$ in the first quadrant, $\theta = 69.06°$. Finally, we express this vector result as

$$R = 149.13 \text{ lb at } 69.06°$$

This is how a vector sum is expressed in **resultant** form.

2.5 SUMMARY

Let us now summarize the procedure for solving for vector sums. This procedure can be organized into seven steps:

1. Draw a point–force diagram.
2. Label all forces (1, 2, 3, etc.)
3. Solve for F_x and F_y for each force.
4. Create a chart to compute ΣF_x and ΣF_y (component form).
5. Solve for $\Sigma F_T = \sqrt{\Sigma F_x^2 + \Sigma F_y^2}$.
6. Determine the resultant quadrant and select the appropriate formula for θ.
7. (a) Find α using

$$\alpha = \tan^{-1}\left|\frac{\Sigma F_y}{\Sigma F_x}\right|$$

   (b) Solve for θ using the formula found in step 6.
   (c) Express your final answer as

$$R = F_T \text{ at } \theta \text{ (resultant form)}$$

The purpose for learning this procedure is so that we might determine how to place a given system into static equilibrium, the condition in which a system exists when it is completely at rest. Now that we understand the concept of vector forces, we can establish the first precept of static equilibrium:

Here are two examples of angled supports that have been used in construction. The forces applied along these angled members must be accounted for by use of vector components and resultants. These structural elements provide additional stability to their structures and enhance their overall structural integrity. *Photo at left courtesy of MML Construction.*

Condition 1: For a system to be in static equilibrium, the sum of all forces internal and external to the system must be zero.

Formula: $\Sigma F_{int} + \Sigma F_{ext} = 0$

If we find, as in the previous example, that the vector resultant of all forces on a system is not equal to zero, we can conclude that the system is not in equilibrium, and must be stabilized. This is done in the following manner. First, construct a **force triangle**, which represents the vector sum in resultant form as shown in Figure 2.8. The stabilizing force for this scenario will have the same magnitude as the sum of the acting forces, but will act in the opposite direction. The direction of the acting force is found by adding 180° to any angle less than 180° or by subtracting 180° from any angle greater than 180°. In this case the angle θ = 69.06° is less than 180°, therefore, the direction of the reacting force is found by 69.06° + 180° = 249.06°. The **stabilizing** force for this example would be expressed as

$$F_{STAB} = 149.13 \text{ lb at } 249.06°$$

Now let us investigate some examples of the use of vectors and vector sums in practical application.

FIGURE 2.8

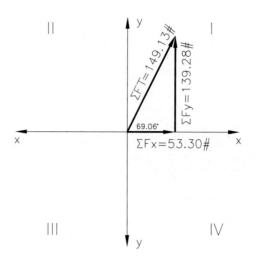

Sample Exercise 2.1

A guy line is fastened to the ground using a cleat with four bolts. If the tensile strength of the cable is 10,000 lb, find the breaking strength and tensile strength of the bolts used to fasten the cleat to the ground if the guy wire is set at an angle of 65° with the ground.

Our first step is to draw a pictorial diagram representing the problem. This is shown in Figure 2.9. We will assume a maximum force of 10,000 lb on the cleat, since that is the value of the tensile strength of the line to which it is attached. Next, we need to construct a point-force diagram so that we can solve for the components of this vector. At present we see that the vector F_T is 10,000 lb or 10 kips at 115°, as shown in our diagrams.

FIGURE 2.9

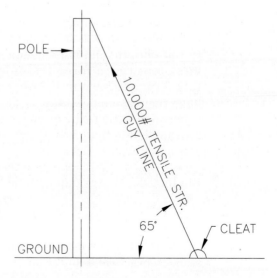

FIGURE 2.10

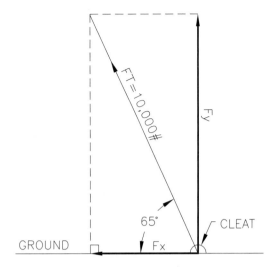

Referring to Figure 2.10 we can see the x and y components of the 10,000-lb vector. We are now able to solve for the components of this vector:

$$F_x = 10{,}000 \text{ lb} \times \cos 65°$$
$$= 4226.18 \text{ lb}$$

and

$$F_y = 10{,}000 \text{ lb} \times \sin 65°$$
$$= 9063.08 \text{ lb}$$

(*Note:* Our sketch shows the x component of this vector to be in the second quadrant, which makes it negative in value. However, since we are only concerned with the strength of the bolts in this problem, the sign of the forces can be ignored.)

The vertical component represents the tension on the bolts, while the horizontal component indicates the shear on the bolts. This is illustrated in Figure 2.11. Since there are four bolts fastening the cleat in place, we assume equal distribution of the forces among them and compute the required bolt strengths by dividing each component by four. We then find

Tensile strength of bolts:

$$\frac{F_y}{4} = \frac{9063.08 \text{ lb}}{4}$$
$$= 2265.77 \text{ lb}$$

FIGURE 2.11

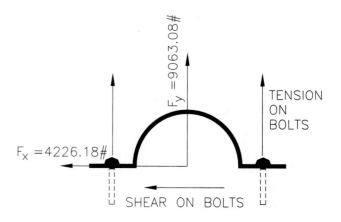

Breaking strength of bolts:

$$\frac{F_x}{4} = \frac{4226.18 \text{ lb}}{4}$$
$$= 1056.55 \text{ lb}$$

Now let's take a look at a problem that requires the use of a vector sum.

Sample Exercise 2.2

A centrifuge wheel is attached to its hub by four evenly spaced spokes. During full speed the centrifuge may generate a centrifugal force of up to 100 lb. It has an additional load capacity of 250 lb at full speed. If the additional load is placed on two opposing spokes, compute the following values:

(a) Tension on the loaded spokes
(b) Tension on the unloaded spokes
(c) Minimum and maximum shear on the hub
(d) The sum of all forces on the hub

Again, we begin this problem by drawing a picture as shown in **Figure 2.12**. Notice that all of the forces in this problem lie along one of the two axes. That makes this problem particularly easy. Notice also that the opposing forces along each axis are equal in value and opposite in direction. As we learned earlier, these are the conditions for stabilizing forces. To find the vector sum of this problem, we pretend to take a snapshot of the wheel at any instant in its motion. We then add the opposing forces to find the tension on the spokes.

(a) The tension on the two loaded spokes is

$$\Sigma |F_x| = 100 \text{ lb} + |-100 \text{ lb}| + \frac{250 \text{ lb}}{2} + \left|\frac{-250 \text{ lb}}{2}\right| = 450 \text{ lb}$$

Analysis of Vectors in Two Dimensions

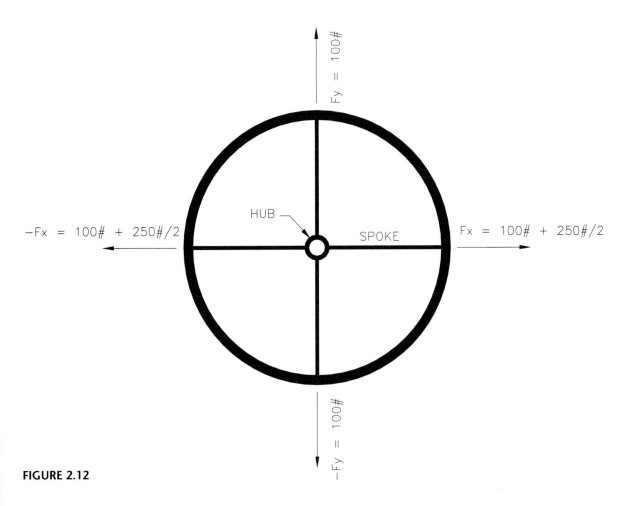

FIGURE 2.12

The sum of the absolute values is used because opposing forces at any point are additive when causing shear at that point. (This is also true of tension and compression.) Therefore, the tension on each loaded spoke is one-half the total load along the x axis or 225 lb.

(b) The tension on the unloaded spokes is 100 lb each, which is equal to F_y in each direction.

(c) The minimum shear on the hub takes place at the points of least loading. This would be along the y axis in this diagram. Therefore, the minimum shear is the sum of all shear forces at the intersection of the y axis with the hub. Again,

$$\Sigma |F_y| = 100 \text{ lb} + |-100 \text{ lb}| = 200 \text{ lb shear}$$

Likewise, the maximum shear will occur at the point of the greatest load on the hub which is at the intersection of the hub with the x axis. So,

$$\Sigma |F_x| = 100 \text{ lb} + |-100 \text{ lb}| + \frac{250 \text{ lb}}{2} + \left|-\frac{250 \text{ lb}}{2}\right| = 450 \text{ lb shear}$$

FIGURE 2.13

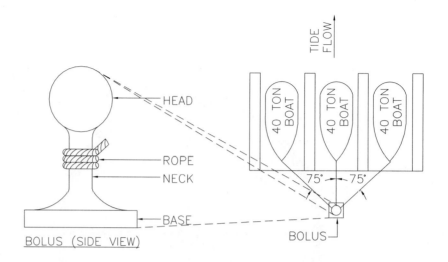

(d) Since the sum of the forces along each of the axes is zero (because equal and opposite forces exist along each axis), there is no vector sum in this problem. This system is in equilibrium but not in static equilibrium since it is in motion.

Sample Exercise 2.3

An iron bolus (shown in Figure 2.13) is used to moor three boats, weighing 40 tons each, to a dock. The boats are located in adjacent slips as shown in Figure 2.13. Perform a structural analysis on the bolus and its fasteners.

Figure 2.14 shows the maximum tension being exerted on the bolus by the mooring ropes. Notice that all three ropes have components that pull along the (+)y axis, making that direction the line of the greatest force. The bolus will experience shear at its base where it is fastened to the pier, and it will experience tension along its neck as it tries to stretch toward the boats. Since the sum of the three forces acting toward the boats (+y) is greater than either x component of the two angled forces, we will use ΣF_y as our maximum force on the bolus.

FIGURE 2.14

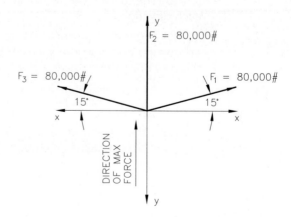

To compute ΣF_y we can simply compute each individual F_y and find their sum:

$$F_{y1} = 80{,}000 \text{ lb sin } 15° = 20{,}705.52 \text{ lb}$$
$$F_{y2} = 80{,}000 \text{ lb sin } 90° = 80{,}000.00 \text{ lb}$$
$$F_{y3} = 80{,}000 \text{ lb sin } 15° = 20{,}705.52 \text{ lb}$$
$$\Sigma F_y = 121{,}411.04 \text{ lb}$$

The result represents the greatest resultant force that can occur in any single direction in the bolus. Therefore, we use this value to assign the appropriate strengths to this system:

$$\text{Tensile strength} = 121{,}411.04 \text{ lb}$$
$$\text{Breaking strength at base} = 121{,}411.04 \text{ lb}$$

The examples given in this chapter should help you solve the following problems related to vectors and their sums.

Note on vector notation:
One way to visually represent the magnitude of a vector is to make the tail of the ray that represents the vector proportionally longer or shorter based on the value of the vector's magnitude. For example, if you drew a 100-lb force using a vector with a 1-in., tail, then you would use a 2-in. tail to represent a 200-lb force. When the vector magnitudes in a problem cover a wide range it is acceptable to estimate the vector lengths while making vectors with smaller magnitudes shorter relative to vectors with larger magnitudes.

This mechanical lift holds its human payload in place by using a number of angled forces. These forces combine to hold the contents of the lift stationary or to move it in various directions. Can you identify the axes along which this piece of equipment operates? *Courtesy of MML Construction.*

Problems

In Problems 1–4 find ΣF_x, ΣF_y, or ΣF_T. Identify each scenario as additive or negating.

1.

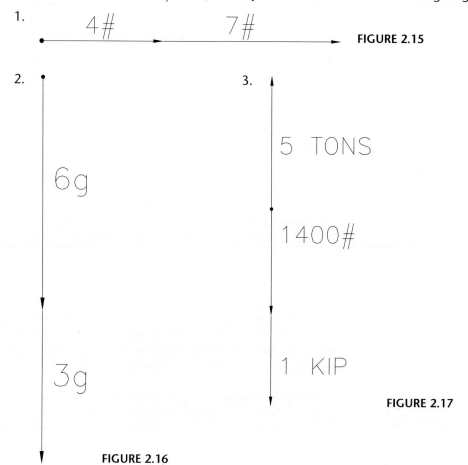

FIGURE 2.15

FIGURE 2.16

FIGURE 2.17

4.

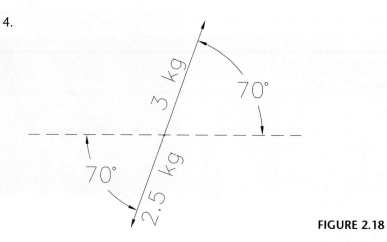

FIGURE 2.18

5. Give the magnitude and direction of each vector below:

(a)

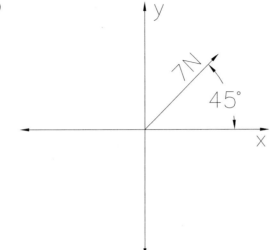

FIGURE 2.19

(b)

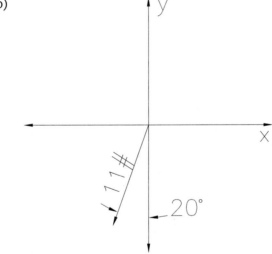

FIGURE 2.20

(c)

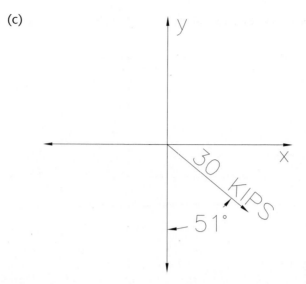

FIGURE 2.21

(d)

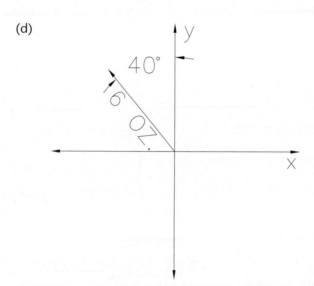

FIGURE 2.22

6. Find the components for each vector in Problem 5.

7. Find ΣF_x and ΣF_y for each scenario:

(a)

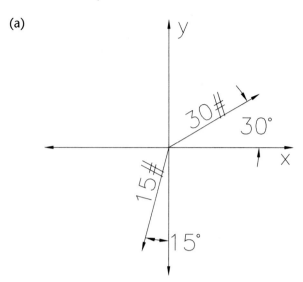

FIGURE 2.23

(b)

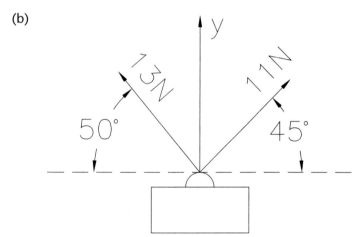

FIGURE 2.24

(c)

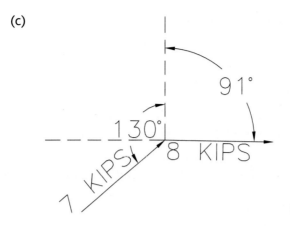

FIGURE 2.25

(d)

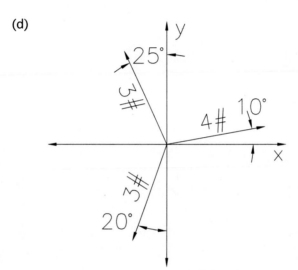

FIGURE 2.26

8. Find the vector resultant ($R = F_T$ at θ) for each scenario in Problem 7.

9. A 400-lb flag pole is anchored into a round concrete base. Wind forces on the flag and pole can cause the pole to sway up to 5° from vertical, creating a downward force of 3.125 kips on the base and an upward force of 1290 lb on the base. Assuming that all forces act perpendicular to the pole's alignment, perform a structural analysis on this system.

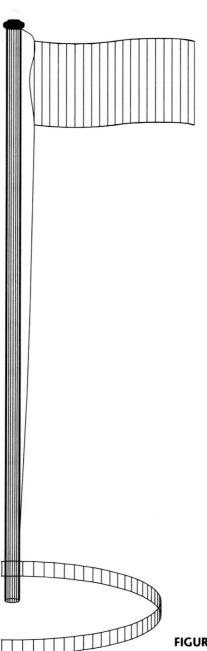

FIGURE 2.27

CHAPTER 3

Work and Energy

3.1 INTRODUCTION

In this chapter we investigate the following items as they relate to work and power:

1. Vertical motion scenarios
   (a) Single force/constant distance
   (b) Single force/multiple distance
   (c) Multiple force/constant distance
2. Potential energy
3. Kinetic energy
4. Kinetic energy loss and transfer
5. Breaking strengths of materials

3.2 WORK DEFINED

Work is defined as a force acting through a distance. The basic formula used to compute the amount of work being done is

$$W = Fd$$

where F = force (in lb or kg), and d = distance traveled (in ft or m). Since work is the product of a force acting through a distance, the units of work are ft-lb (English), and kg-m (S.I.). Other units include the inch-pound (in-lb) and the gram-centimeter (g-cm). A **foot-pound** (ft-lb) is the amount of work required to raise a one-pound object to a height of one foot above its original position.

To illustrate this concept we investigate the amount of work required to raise a 30-lb object to a height of 3 ft from its original position:

$$W = Fd \quad F = 30 \text{ lb}, d = 3 \text{ ft}$$
$$= 30 \text{ lb} \times 3 \text{ ft}$$
$$= 90 \text{ ft-lb}$$

The object gained work, or was worked upon, because the value of the work was positive. We can likewise investigate the result of lowering the same object 2 ft below its original position:

$$W = Fd \quad F = 30 \text{ lb}, d = -2 \text{ ft}$$
$$= 30 \text{ lb} \times (-2 \text{ ft})$$
$$= -60 \text{ ft-lb}$$

Remember from our study of vectors in Chapter 2 that motion upward is positive, and motion downward is negative. In this instance the object lost work, or we can say that it performed work on its surroundings.

These two concepts of work will be used in the ensuing sections as we discuss additional work scenarios and energy in systems. Notice that the examples given thus far involve vertical motion. We will continue to study the effects that vertical motion of objects has on work produced by those objects. Later in this chapter we will also look at systems having lateral (horizontal) motion.

3.3 ADDITIONAL WORK SCENARIOS

Single Force/Multiple Distance

The two previous illustrations are examples of the single force/constant distance scenario of work. In this scenario one object moves through one distance producing one work result. This is the simplest of all the work scenarios. Next we will look at the single force/multiple distance scenario.

To illustrate this concept we will take a 16-lb bowling ball and move it up three 1-ft-high steps, as shown in Figure 3.1. The force F will be the weight of the ball. This force remains constant regardless of where we move the ball and regardless of how many times we move the ball. Therefore, we are dealing with a single force only in this example. Notice, however, that the ball must be moved up three steps. Each time we move the ball up another step it is moved an additional distance. In moving the ball up the first step we perform 16 ft-lb of work ($W = 16 \text{ lb} \times 1 \text{ ft}$). This first distance traveled in the y direction we will call d_1. The work needed to move the ball through the distance d_1 is then

$$W_1 = Fd_1$$
$$= 16 \text{ lb} \times 1 \text{ ft} = 16 \text{ ft-lb}$$

FIGURE 3.1

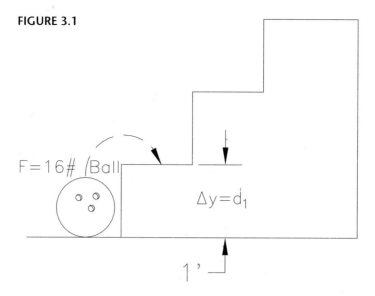

To move the ball from the first step to the second requires the addition of another 16 ft-lb of work, as shown in Figure 3.2. The sum of the work done on the ball thus far is equal to the weight of the ball multiplied by the total distance it has been moved, so that

$$\Sigma W_{1\text{-}2} = F(d_1 + d_2)$$
$$= F\Sigma d$$

This then is the formula for the single force/multiple distance scenario. Using this new formula we find that the work necessary to move the ball up two steps is

$$\Sigma W_{1\text{-}2} = 16 \text{ ft} \times (1 \text{ ft} + 1 \text{ ft})$$
$$= 32 \text{ ft-lb}$$

FIGURE 3.2

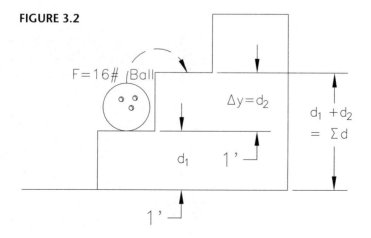

FIGURE 3.3

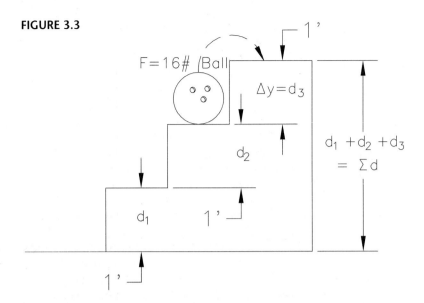

Finally, we will move the ball up to the third step, adding yet another 16 ft-lb of work to the ball. This is shown in Figure 3.3. Again we will apply the formula above, but this time we will add all of the distances from all three steps together:

$$\Sigma d = d_1 + d_2 + d_3 \\ = 1 \text{ ft} + 1 \text{ ft} + 1 \text{ ft} \\ = 3 \text{ ft}$$

Since we already know that $F = 16$ lb we can now complete this problem:

$$\Sigma W_{1\text{-}3} = 16 \text{ lb} \times 3 \text{ ft} \\ = 48 \text{ ft-lb}$$

This vehicle is traveling down a hill. The amount of work done by the vehicle as it moves down the hill can be expressed as work components using the same method discussed in Chapter 2 for force or vector components. The amount of work done vertically by the vehicle as it moves down the hill is the product of its weight and the vertical distance it travels, regardless of the length of the road or the angle of the slope. The total work done by the vehicle is a function of the actual distance traveled along the line of force (road).

Multiple Force/Constant Distance

The last work scenario that we will discuss is the multiple force/constant distance scenario. In this scenario we have two or more forces acting along the same distance. To illustrate let us consider a crane that returns 4 times to lift objects varying in weight to the top of a 10-story building. We assume that the weight of each of the objects is as follows: (1) a 1000-lb bundle of steel rods, (2) a square of 60-lb shingles stacked 10 high weighing a total of 1800 lb, (3) 500 lb in roofing materials, and (4) a ton of tools and equipment.

To raise the first load, which we will call F_1, to the top of the building requires

$$W = 1000 \text{ lb} \times 100 \text{ ft}$$

or 100,000 ft-lb of work. This is illustrated in Figure 3.4. Since the crane has now done this amount of work in the first stage of this scenario, we will remember this value and keep a running tally of the total work done by the crane in the upper left corner of the next three figures. This tally will reflect only the work done on the loads currently shown on the roof of the building.

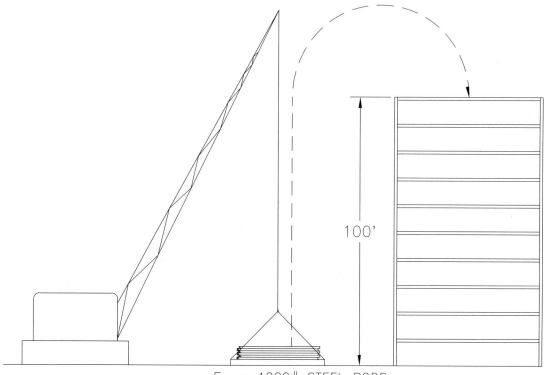

FIGURE 3.4

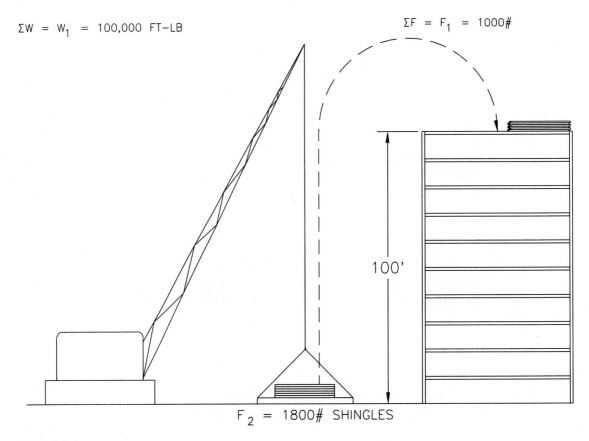

FIGURE 3.5

To raise the 1800 lb of shingles (F_2) to the roof (Figure 3.5) will require

$$W = 1800 \text{ lb} \times 100 \text{ ft}$$

or, an additional 180,000 ft-lb of work. Our total work in the problem, shown in Figure 3.6, now stands at

$$\Sigma W_{1\text{-}2} = 100{,}000 \text{ ft-lb} + 180{,}000 \text{ ft-lb} = 280{,}000 \text{ ft-lb}$$

Now the crane lifts the 500-lb load of roofing materials, which we will call F_3, to the roof of the building. This requires the crane to do another 50,000 ft-lb of work, bringing the total work done, or $\Sigma W_{1\text{-}3}$ to 330,000 ft-lb. Finally, the last load of 2000 lb, shown in Figure 3.7, is lifted to the roof. This requires another 200,000 ft-lb of work to be done. So, in total, the crane does 530,000 ft-lb of work.

Since the sum of the weight of all of the objects was lifted over the same distance, we can find their sum and multiply it by the common distance traveled to find the total work in this scenario. This is expressed mathematically as

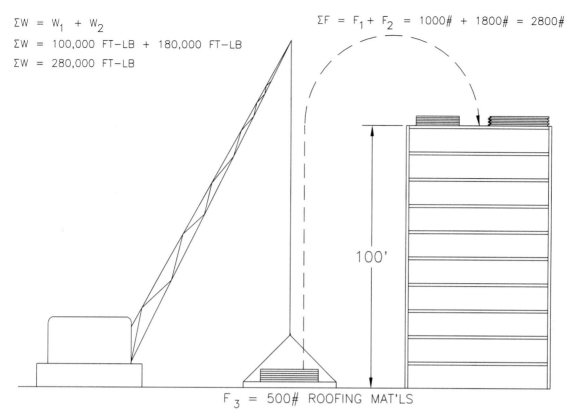

FIGURE 3.6

$$\Sigma W_{1\text{-}4} = (F_1 + F_2 + F_3 + F_4)d$$

or

$$\Sigma W_{1\text{-}4} = \Sigma F d$$

The general formula for the multiple force/constant distance scenario is

$$\Sigma W = \Sigma F d$$

Applying this formula to the above problem yields

$$\begin{aligned}\Sigma W &= (1000 \text{ lb} + 1800 \text{ lb} + 500 \text{ lb} + 2000 \text{ lb}) \times (100 \text{ ft}) \\ &= (5300 \text{ lb}) \times (100 \text{ ft}) \\ &= 530{,}000 \text{ ft-lb}\end{aligned}$$

Multiple Force/Multiple Distance

There is also a fourth scenario, known as the multiple force/multiple distance scenario, which we will not explore due to its complex nature. However, the student who is familiar with the three scenarios discussed previously in this chapter should be able to apply these principles to any work scenario where

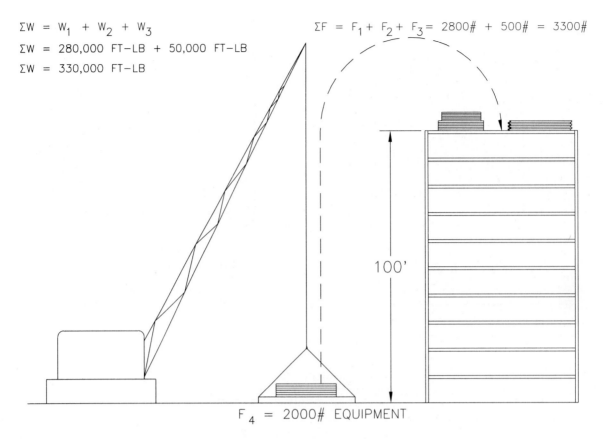

FIGURE 3.7

more than one force and/or distance are concerned. A general rule of thumb is to treat all work problems as if they are made up of one or more individual problems. Then each problem can be approached using the concepts covered in this chapter.

3.4 POTENTIAL ENERGY

Potential energy is defined as a type of stored energy and is measured as a function of an object's position relative to the ground The potential energy of an object is computed by multiplying the weight of the object by its distance above the ground in feet. The formula for calculating potential energy (PE) is

$$PE = Fd$$

Notice that this is the same formula that we used to calculate work earlier. We can conclude, then, that the potential energy of an object is a direct result of the work added to or taken from that object.

Industrial equipment, such as the crane and the lift shown in these photographs, perform work on various loads. The lift, for example, is being used to raise a fixed number of people to various heights to work on or inspect a part of a building. This illustrates the single force/multiple distance work scenario. The crane shown in the other photograph is being used to remove several different loads from the flatbed. This is an example of the multiple force/constant distance scenario. Photo at right courtesy of MML Construction.

If we consider the bowling ball in Figure 3.1 we find that its initial PE is

$$PE_I = 16 \text{ lb} \times 0 \text{ ft} = 0 \text{ ft-lb}$$

Since the ball is resting on the ground, it has no distance above the ground and, therefore, no PE relative to the ground at this point. If we move the ball up to the next step, then we have performed 16 ft-lb of work on the ball and have added 16 ft-lb of potential energy to the ball. Whenever work is done on or by a system, this brings about a change in potential energy for that system. The change in the potential energy (ΔPE) of a system is always a direct result of the work being done on or by the system.

In Figure 3.2 the ball has been moved up 1 ft and therefore

$$\Delta PE_{0\text{-}1} = F \Delta d$$
$$= 16 \text{ lb} \times 1 \text{ ft} = 16 \text{ ft-lb}$$

Since the ball has moved upward, the change in PE is positive and energy has been added to the system. Therefore, the change in PE for an object moving upward indicates the amount of energy gained by the system. Likewise, an object moving downward would lose potential energy because the work would be done by the system rather than on the system.

To find the final PE (PE_T) of an object relative to its original position we use the following formula

$$PE_T = PE_I + \Delta PE$$

If the object gains PE, then $PE_T > PE_I$, whereas if the object loses PE, then $PE_T < PE_I$. The final potential energy for the bowling ball after it has been raised 1ft is:

$$PE_T = 0 \text{ ft-lb} + 16 \text{ ft-lb} = 16 \text{ ft-lb}$$

This answer indicates a net gain in potential energy.

The scenarios given in this chapter relating to work are also applicable to energy. Energy may be gained or lost by an object moving over several distances (constant force/multiple distance) or several objects moved over the same distance may collectively gain or lose energy (multiple force/constant distance). Let us again look at the bowling ball, but this time we will begin with the ball at the top of the steps, as shown in Figure 3.8. Notice that in this case the ball is traveling downward, which will result in a net loss of potential energy. We can calculate that loss by examining this problem using the constant force/multiple distance scenario. In this case, the constant force of 16 lb moves down three steps of 1 ft each, losing 16 ft-lb at each step (ΔPE

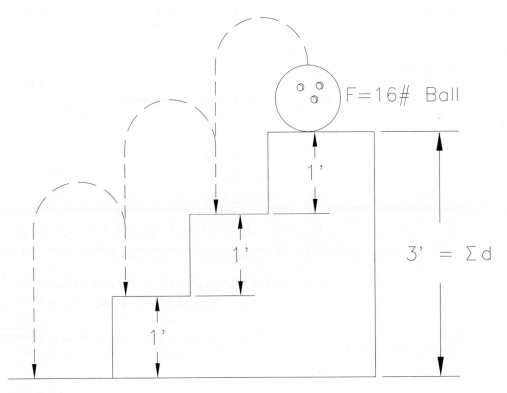

FIGURE 3.8

= −16 ft-lb). Since there are three steps, we simply add all of the distances through which the ball travels and multiply that result by the weight of the ball. Hence,

$$\Delta PE_{3\text{-}0} = F\, \Sigma d$$
$$= 16\text{ lb} \times [(-1\text{ ft}) + (-1\text{ ft}) + (-1\text{ ft})]$$
$$= 16\text{ lb} \times (-3\text{ ft}) = -48\text{ ft-lb}$$

So the ball loses 48 ft-lb of potential energy as it falls down the steps.

3.5 KINETIC ENERGY

The **law of conservation of energy** states that energy can be neither created nor destroyed, but can be transformed. Examples of this can be found in a light bulb, which converts electrical energy to light and heat; an automobile engine, which transfers fuel energy into heat, friction, and motion; and your body, which transfers chemical energy from the foods you eat to heat and motion as you move and work. Likewise, when potential energy is lost from an object it must go somewhere as another form of energy. We call this type of energy **kinetic energy**, because it is an energy transferred from one medium to another as a result of motion. Kinetic energy is defined as a type of imparted (released) energy and is measured as a function of the velocity of an object at impact with a surface.

To illustrate the effects of kinetic energy (KE) we can take a hammer and drop it onto a cube of ice from a height of 1 in. Depending on the size of the ice cube, there may be little damage done. If we take the same hammer, however, and drop it from a height of 6 ft, it will do considerably more damage. Why? Because the hammer has a greater velocity when it falls 6 ft than it does when it falls 1 in. Notice that a 3-lb hammer resting 1 in. above the ground has only 0.25 ft-lb of potential energy stored in it, whereas the same hammer resting 6 ft above the ground has 18 ft-lb of energy stored in it.

We can see then that PE and KE are related in some fashion. In fact, they are related in the following way:

$$KE = -\Delta PE$$

The kinetic energy an object imparts on a surface at impact by falling from a distance has the same value as the potential energy it has lost in falling through that distance. Notice that if we roll the bowling ball off the right side of the top step in Figure 3.8, the ball will lose 48 ft-lb of PE in the fall. This energy will be transferred to the ground as KE at impact.

Effective Falling Distance

The above principle works great if you know how far an object has fallen, but what if you don't know how far it fell? What if it has a velocity, like a car or bicycle, but does not fall at all? Of course, such objects have energy and would release this energy if they were to impact upon a surface. An object traveling horizontally has no falling distance d. We can, however, compute the effective falling distance of an object given its velocity by the formula below:

$$d = \frac{v^2}{2g}$$

Since kinetic energy occurs as a result of the velocity of an object striking another surface, we can find the energy an object imparts when it strikes a surface if we know its weight and its velocity. The velocity is measured in **feet per second** (fps) and the value $g = 32.2$ ft/s². This is the constant of gravitational acceleration for the earth. The value of the effective falling distance d, which is found using the formula above, represents the height from which an object would need to fall in a vacuum in order to reach the velocity v.

If we substitute the above formula for d in the formula KE = $F \times$ d we obtain

$$\text{KE} = \frac{Fv^2}{2g}$$

To illustrate the use of this formula we will investigate the effect that a bullet shot out of a gun has on a piece of sheet metal as the bullet passes through it. Assuming that the bullet has a weight of 0.02 lb and is traveling at 1150 ft/s (Figure 3.9), we can compute its initial PE using the formula given in this section. Substituting the appropriate values we have

$$\text{PE}_\text{I} = \frac{0.02 \text{ lb} \times (1150 \text{ fps})^2}{2 \times 32.2 \text{ ft/s}^2}$$

Solving for PE_I we find that $\text{PE}_\text{I} = 410.714$ ft-lb. If this bullet strikes a sheet metal plate and loses 55% of its energy at impact, we can investigate the following:

FIGURE 3.9

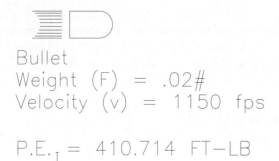

FIGURE 3.10

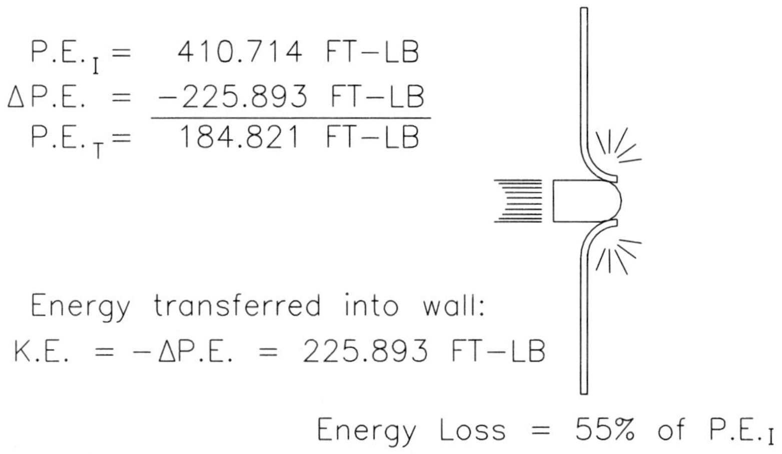

$$P.E._I = 410.714 \text{ FT-LB}$$
$$\Delta P.E. = -225.893 \text{ FT-LB}$$
$$P.E._T = 184.821 \text{ FT-LB}$$

Energy transferred into wall:
$$K.E. = -\Delta P.E. = 225.893 \text{ FT-LB}$$

Energy Loss = 55% of $P.E._I$

(a) Energy transferred to plate at impact (KE)
(b) Energy remaining in bullet after impact (PE_T)
(c) Velocity of the bullet after impact (v_T)
(d) Breaking strength of the metal plate (BkStr in ft-lb/in²)

Solution:

(a) To calculate the amount of energy transferred to the plate at impact, as shown in Figure 3.10, we need to find out how much energy was lost by the bullet at impact with the plate. To do this we compute the energy loss value using the percent energy loss given in the problem:

$$\Delta PE = -(410.714 \text{ ft-lb} \times 0.55)$$
$$= -225.893 \text{ ft-lb}$$

(b) Computing the remaining energy in the bullet is easy at this point. We simply subtract the loss in PE from the initial PE and we find that

$$PE_T = 184.821 \text{ ft-lb}$$

(c) To calculate the velocity of the bullet after impact we will have to manipulate the energy equation used in finding PE_I to isolate the velocity variable v. Let us first rewrite this equation to reflect the current conditions involving final PE and final velocity:

$$PE_T = \frac{Fv_T^2}{2g}$$

Solving for v_T we now find

$$v_T = \sqrt{\frac{2g PE_T}{F}}$$

Parking posts are designed with adequate breaking strength to be able to resist or reduce the forces of moving objects. The posts are fashioned so that they are able to absorb a certain amount of the kinetic energy that a moving vehicle can impart. These parking posts are located around a power line pole anchor to protect it from vehicular impact.

Using this formula we can substitute $PE_T = 184.821$ ft-lb and $F = 0.02$ lb to obtain

$$v_T = \sqrt{\frac{2 \times 32.2 \text{ ft/s}^2 \times 184.821 \text{ ft-lb}}{0.02 \text{ lb}}}$$

$$= 771.90 \text{ fps}$$

By the way, if we needed to find the above velocity in mph we could use the conversion factor 1 mph = 1.47 fps, so that

$$v_T = 771.44 \text{ fps} \div 1.47 \text{ fps/mph}$$
$$= 524.79 \text{ mph}$$

(d) Finally, to determine the breaking strength of the metal plate we must first find out how much energy is required to break the plate at the point of impact. This is equal to the bullet's loss of energy at the instant of impact with the plate. From Figure 3.10 we see that this value is 225.893 ft-lb. Next we need to know the area of the surface struck by the bullet. If we assume that the bullet has a $\frac{1}{8}$-in. diameter then the area struck (A) becomes

$$A = \pi r^2$$
$$= \pi(\tfrac{1}{16} \text{ in.}^2)$$
$$= 0.0123 \text{ in.}^2$$

Shear may occur as a result of an object moving with a certain velocity colliding with a surface. In such instances the breaking strength of a material may be exceeded, resulting in the condition shown in this photograph. Concrete has a relatively high breaking strength, so, it would have taken a large amount of energy focused on a small area of this barrier in order to break it.

Last of all we need to determine the breaking strength value of the plate by dividing the KE transferred at impact by the area struck. Then, we have

$$\text{BkStr} = \frac{-\Delta \text{PE}}{A}$$

so

$$\text{BkStr} = \frac{-(-225.893 \text{ ft-lb})}{0.0123 \text{in.}^2}$$

Therefore,

$$\text{BkStr} = 18{,}365.29 \text{ ft-lb/in.}^2$$

3.6 SUMMARY

We can now establish the second precept of static equilibrium based on the material covered in this chapter. Since work is the result of a force acting through a distance we know that both a force and a distance must exist for work to occur. If the first precept for static equilibrium is true then the sum of all forces is zero. Since there is no total force, there can be no total work (anything multiplied by zero is zero) We, therefore, have the following:

Condition 2: For a system to be in static equilibrium, the sum of all work done on and by the system must be zero.

Formula: $\Sigma W_{\text{int}} + \Sigma W_{\text{int}} = 0$

Now let us look at some sample exercises to illustrate the concepts of work and energy covered in this chapter.

Sample Exercise 3.1

A 3-ton pile driver is raised 3 ft above a piling before it is released. Find the following information related to the pile driver:

(a) Its maximum potential energy
(b) The kinetic energy with which it strikes the piling
(c) The total work required to strike the piling 10 times

Referring to Figure 3.11 we can see the setup for this problem.

(a) To find the maximum PE of the pile driver all we need to do is determine the highest point that it will reach during its course of motion. Since the PE of an object is related to its height above the ground, the highest point will yield the greatest PE

Since the highest point is 3 ft, this will be our value for d. The value for F is 3 tons, which equals 6000 lb. So,

$$PE_{Max} = PE_T = 6000 \text{ lb} \times 3 \text{ ft} = 18{,}000 \text{ ft-lb}$$

(b) Since the pile driver will fall 3 ft before striking the piling it will lose 18,000 ft-lb of energy. This will be transferred to the piling as KE and will serve to drive it into the ground.

$$\therefore KE = -(-18{,}000 \text{ ft-lb}) = 18{,}000 \text{ ft-lb}$$

(c) To find the amount of work needed to strike the piling 10 times we need to know how much work is required to raise the pile driver the same number of times. This becomes a constant force/multiple distance work problem

FIGURE 3.11

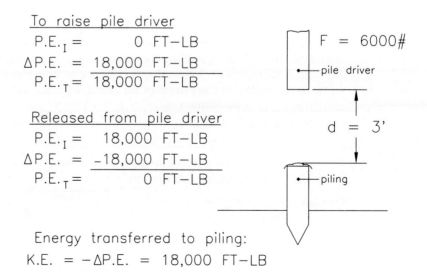

since the 6000-lb pile driver must move over the 3-ft distance 10 times. The solution is

$$\Sigma W = F\Sigma d$$
$$= 6000 \text{ lb} \times (10 \times 3 \text{ ft})$$
$$= 180,000 \text{ ft-lb}$$

Sample Exercise 3.2

A 1.5-ton car traveling at 70 mph strikes a guard rail, tearing out a 1-ft × 1-ft section of the rail as it passes through. The car is still traveling at 20 mph after the collision. Compute the breaking strength of the rail.

To compute the rail's breaking strength, we know that we must first find the KE imparted into the rail at impact and the area struck by the car. The area struck is easy to find. Using the information in the problem we find

$$(\text{Area of the cross-section struck}) \; A_{cs} = 12 \text{ in.} \times 72 \text{ in.} = 864 \text{ in.}^2$$

To find the KE imparted at impact we must find PE_I and PE_T so that we can find ΔKE at impact with the rail. Using the velocity formula for energy we find:

$$PE_I = \frac{3000 \times (70 \times 1.47)^2}{2(32.2)}$$
$$= 493,248.91 \text{ ft-lb}$$

and

$$PE_T = \frac{3000 \times (20 \times 1.47)^2}{2(32.2)}$$
$$= 40,265.22 \text{ ft-lb}$$

So

$$KE = -\Delta PE = -(40,265.22 \text{ ft-lb} - 493,248.91 \text{ ft-lb})$$
$$= 452,983.7 \text{ ft-lb}$$

Therefore,

$$BkStr = 452,983.7 \text{ ft-lb}/864 \text{ in.}^2$$
$$= 524.29 \text{ ft-lb/in.}^2$$

Sample Exercise 3.3

An aircraft weighing 64,400 lb produces 200,000 ft-lb of thrust at takeoff. It climbs at an angle of 12°. Find the following values at takeoff:

(a) Air speed in mph
(b) Ground speed in mph
(c) Rate of ascent in fps

FIGURE 3.12

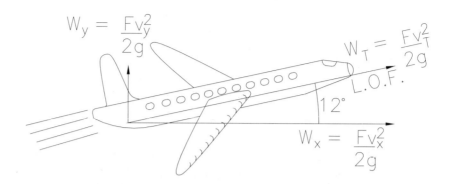

(a) Air speed is defined as the speed at which a plane is traveling along its line of motion/force. Factors such as wind velocity, air pressure, and angle of trajectory will have an effect on the speed at which a plane travels through the air. If we look at Figure 3.12 we will see air speed is noted by v_T. To find air speed we will use $F = 64{,}400$ lb (weight of the plane) and $W_T = 200{,}000$ ft-lb (thrust along the line of motion/force). To find the velocity of the plane we use the formula for energy due to velocity:

$$v_T = \sqrt{\frac{2gW_T}{F}}$$

In this equation W has been substituted for energy because the work being done by the engine thrust on the plane is the motivation that causes the plane to move forward. This is an illustration of Newton's third law of motion.

Now, substituting the given values into the equation we find

$$v_T = \sqrt{\frac{2(32.2 \text{ ft/s}^2)(200{,}000 \text{ ft-lb})}{64{,}400 \text{ lb}}}$$

$$= 200 \text{ fps} \div 1.47 \text{ fps/mph} = 136 \text{ mph}$$

(b) Next, we'll find the ground speed of the plane. This is the actual speed at which the plane moves from point to point relative to points located on the ground. In Figure 3.12 this motion occurs along the x axis. This velocity (v_x) is a component of the air speed (v_T) and is computed using the force triangle in Figure 3.13, where $v_x = v_T \cos \alpha$. We then find that

$$v_x = 136 \text{ mph } (\cos 12°) = 133 \text{ mph}$$

FIGURE 3.13

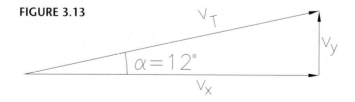

(c) Finally, the rate of climb (v_y) can be found by the formula $v_y = v_T \sin \alpha$, which yields

$$v_y = 200 \text{ fps } (\sin 12°) = 42 \text{ fps}$$

Problems

1. Compute the total work done in each situation given below:

   (a) A 200-lb man climbs to the top of an 8-ft-tall ladder.
   (b) A car climbs a 40-ft high mound, then a 200-ft high hill, and finally a 2500-ft high mountain.
   (c) An elevator lifts four 190-lb people, a 100-lb box, and a 60-lb crate to a height of 20 ft.
   (d) A 15-lb cat falls out of a 60-ft tall tree.

   In Problems 2–4, compute (a) the initial potential energy of the object (PE_i), (b) the change in potential energy of the object (ΔPE), and (c) the final potential energy of the object (PE_T). Express all final answers in foot-pounds.

2. A crane raises a 2-ton wrecking ball from resting on the ground to a height of 30 ft.

3. The crane in the Problem 2 lowers the same wrecking ball from a height of 30 ft to a height of 20 ft.

4. A weightlifter replaces a pair each of 25-, 30-, and 35-lb dumbbells from the floor onto a 2-ft., 6-in.-high weight rack.

5. Find the KE or ΣKE imparted on the ground for Problems 2–4 if the objects were to fall to the ground from their final resting place in each problem.

6. An object at rest at the top of a 3000-ft cliff falls to a shelf 1750 ft high and stops. Find the kinetic energy imparted by the object at impact with the surface.

7. Compute the effective falling distance of a bicycle traveling at 30 mph.

8. If the bicycle in Problem 7 weighs 25 lb and is carrying a 170-lb rider, how much energy will it impart on a tree branch as the rider passes through it if it loses 5% of its total energy in the collision?

9. A 6-in.-diameter ramrod is used to break through a wall. If the ramrod weighs 70 lb and is swung at a velocity of 30 mph, what is the breaking strength of the wall?

10. A 1.5-ton car is driven along a horizontal distance of one mile. How much work has been done in moving the car vertically if it is being driven

   (a) up a 3% incline
   (b) up a 7% incline
   (c) up a 10% incline
   (d) down a 5% incline

   (*Note:* The percent incline is found by multiplying the ratio of the road rise to the road run by 100.)

11. A 200-lb go-cart and driver start at the top of a hill. At the finish line the odometer in the cart reads 0.3 miles more than it did before the run. If the cart went down a $7\frac{1}{2}$% incline (neglecting friction)

   (a) how fast was it going at the finish line?
   (b) what was its potential energy at the top of the hill?
   (c) what was its potential energy at the finish line?

   Compare your answers in parts b and c. What does this tell you? Why do you think this is so?

12. A 2-ton car strikes a guard rail at 70 mph, imparting 70% of its kinetic energy into the rail. It then continues into a cinder block wall where it comes to a complete stop. Find

   (a) the breaking strength of the guard rail if the area struck by the car is 14 in. high by 6 ft wide.
   (b) the kinetic energy released into the wall.

13. A 1.3-ton vehicle collides with a concrete wall, losing 60% of its kinetic energy at impact. The vehicle strikes an area on the wall $3\frac{1}{2}$ ft by 4 in. If the wall's breaking strength is 1000 ft-lb/in.^{2}, determine the velocity of the car in miles per hour at the following points of its travel:

   (a) at the instant of impact with the wall
   (b) at the instant after impact with the wall

   Compare the change in kinetic energy in this problem to the change in the car's velocity. What do you notice? Why do you think that this is so? Based on this information can you formulate a general rule that relates the change in velocity of an object of unknown mass to its change in potential energy due to that change in velocity?

14. A one-kiloton bomb (1 kiloton = 1000 tons) is detonated midair. Parts of the bomb casing along with the expanding gases fly in all directions about the point of detonation (ground zero). If you were to create an elaborate three-dimensional point–force diagram to represent the weight and velocity of

every particle in this explosion, what result would you expect to find? Why? (*Hint:* Disregard the downward velocity of the bomb at the time of the explosion. Treat the explosion as if it had occurred while the bomb was stationary, floating in space.)

Moments

4.1 INTRODUCTION

In this chapter we investigate the following items as they relate to moments:

1. First-order levers
2. Centroids
3. Second-order levers
4. Angled forces
   (a) Acting
   (b) Stabilizing
5. Stabilizing techniques

4.2 MOMENTS DEFINED

A **moment** is defined as the effect of a force applied along an object at a distance from a fixed point which causes rotation about that fixed point. The value of a moment about a fixed point is found by multiplying the value of a force applied perpendicular to a surface by the distance that the force is from the fixed point. To better understand the concept of moments, let's look at a few terms that we have defined which correspond to key locations in a moments problem.

In Figure 4.1 you will notice four abbreviations:

POI: **point of interest**—the fixed point about which rotation occurs. Also known as the **fulcrum**.

POA: **point of action**—the point along the line of interest at which the external force is applied.

LOI: **line of interest**—the line on which the external force is applied. The POI and the POA are both found along this line.

FIGURE 4.1

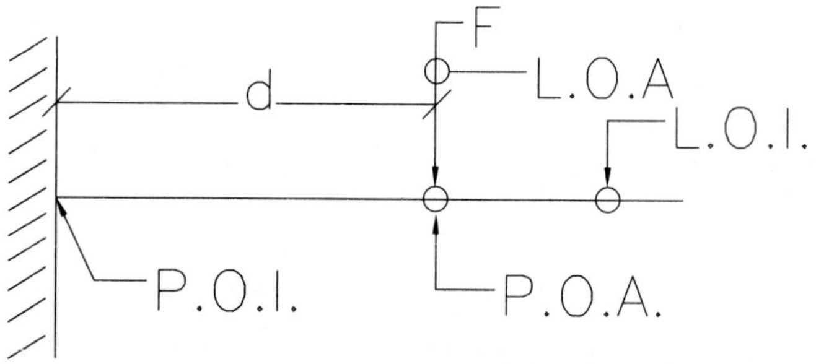

LOA: line of action—the line along which external force is applied. The LOA is always perpendicular to the LOI (LOA ⊥ LOI).

In Figure 4.1, F represents the magnitude of the external force (lb, kg, etc.) and d represents the distance (ft, m, etc.) from the fulcrum, or fixed point, to the force at the POA. As we shall see shortly, the direction of the distance is equally as important as the magnitude of the distance.

4.3 CONVENTION OF MOMENTS

Since moments create rotation, they are not measured according to linear direction, such as up, down, left, or right. Instead, the direction of rotation is defined as being **clockwise (cw)** or **counterclockwise (ccw)**. To mathematically combine moments, a convention of signs must be agreed on. This text bases that convention on the standard directional convention used in the Cartesian coordinate system.

Figure 4.2 illustrates how various forces and their locations along the x axis cause rotation about the origin of the coordinate system. If the x axis were a seesaw attached to the ground at the point (0,0), then the application a sin-

FIGURE 4.2

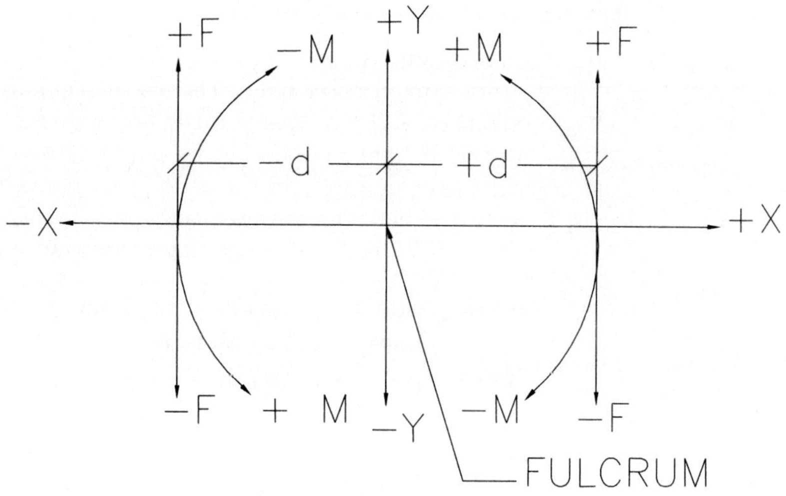

gle force either up or down at a single location to the right or left of the fulcrum would cause the seesaw to rotate either clockwise or counterclockwise.

If we maintain all standard sign conventions throughout, then we will arrive at the four conditions listed below concerning the signs convention of moments:

THE SIGNS CONVENTION OF MOMENTS

If a force is located to the _____ of the fulcrum	And it is applied in a(n) _____ direction	Then it creates a _____ moment
Right (+d)	Upward (+F)	CCW (+M)
Right (+d)	Downward (−F)	CW (−M)
Left (−d)	Upward (+F)	CW (−M)
Left (−d)	Downward (−F)	CCW (+M)

Therefore, clockwise moments are negative and counterclockwise moments are positive.

This children's seesaw is a typical example of a second-order lever (discussed later in this chapter). The fulcrum is located at the center of the lever in this case. Notice the covered springs at the center below the aparatus. These springs provide for more smooth operation of the ride, especially when two children of different weights are using it (see Sample Exercise 4.2 at the end of this chapter).

FIGURE 4.3

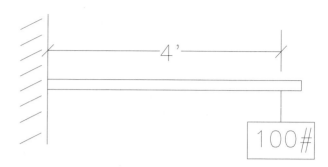

4.4 FIRST-ORDER LEVERS

A first-order lever is defined as a lever with a fulcrum located at one of its ends. The lever itself can be any rigid object, such as a beam, plank, pipe, or post, that is fastened to a fixed surface or supported at one end. To better understand the concept of moments and the roles that they play in structural applications, let us consider the following situation involving a first-order lever.

A streetside store owner wants to hang a 100-lb sign outside her establishment. The sign needs to hang 4 ft away from the wall. What effect will this have on the point where the support is fastened to the wall? Figure 4.3 shows us the system. Obviously, without adequate support at the wall the

These two lampposts appear to be two identical examples of first-order levers, but a closer look reveals that each one is making use of a different external stabilizing force. The lamppost on the left is being stabilized by a rigid body placed in compression on the underside of the lamp structure. The lamp on the right has been designed with a bent line of interest, which allows a pair of threaded rods in tension to stabilize its rotation. In each case the normal component for the stabilizing force prevents rotation and maintains static equilibrium.

FIGURE 4.4

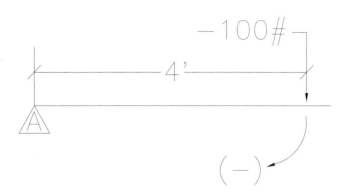

support beam will be pulled downward on its right-hand end, causing clockwise rotation about the fixed point on the left. To determine the actual value of the work being done at that point, seen as point A on the point–force diagram in Figure 4.4, we simply multiply the weight of the sign (100 lb) by the distance it is located from the wall (4 ft):

$$M_A = -100 \text{ lb} \times 4 \text{ ft}$$
$$= -400 \text{ ft-lb}$$

The answer is read, "The moment about point A is negative 400 foot-pounds." This indicates that the moment is clockwise because its sign is negative. The formula then for the moment about a point A due to a force applied at a distance d is:

$$M_A = Fd$$

Notice that the units for moments are the same as the units for work and energy. This is because moments create work at a single point, the fulcrum. Moments also create a twisting effect about the fulcrum known as **torsion** (see Chapter 1). For this reason, the units associated with torsion are the same as those associated with work, energy, and moments.

Unfortunately, real life is not as simple as the one-step problem shown above. Most existing structures have many forces acting upon them, which can cause a variety of moments at any given point on the structure. In such cases we must determine the sum of the moments (ΣM) about a given fixed point.

Upon examination of the system shown in Figure 4.5, we can see that two distinct external forces act upon the support shown. Notice that there is a 2000-lb weight hanging 3 ft from the wall and a 1000-lb spring pushing upward 6 ft from the wall. Both of these forces will create separate moments about the fixed point at the wall. The point–force diagram in Figure 4.6 shows the location, magnitude, and direction of these forces.

FIGURE 4.5

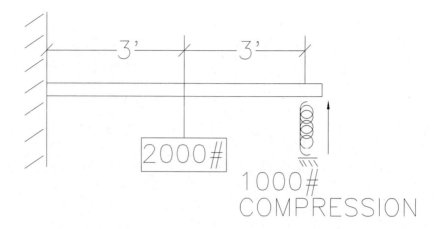

FIGURE 4.6

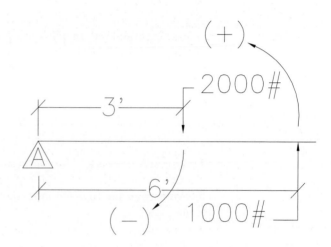

The easiest way to compute the sum of two or more moments is by means of a moments chart. The moments chart for this example is shown as Table 4.1. If we label the forces from the fulcrum outward, then the 2000-lb force becomes force 1 (F_1) and the 1000-lb force becomes force 2 (F_2). Then, we simply fill in the chart as shown and perform the indicated operations. We find that the compressive strength of the spring located at the end of the support is sufficient to counteract the rotation that would have been caused by the 2000-lb weight hanging along the support. This system is, then, in **static equilibrium**.

TABLE 4.1

Force	F (lb)	d (ft)	M_A (ft-lb)
1	−2000	3	−6000
2	+1000	6	+6000

$\Sigma M_A = 0$ ft-lb

This four-person seesaw uses compression springs placed off-center to counteract the moments caused by the children who are playing on it. The children are the acting forces, while the springs are the reacting forces. Each causes a moment to occur about the center of the ride.

4.5 CENTROIDS

As we have seen, forces may act on a surface at a single point, causing rotation about another point. However, not all forces are designed to act at a single point. Take, for example, the 50-lb box shown in Figure 4.7. The box is 2 ft wide and, therefore, does not act at a single point along the plank. How do we locate this force on a point–force diagram?

Physicists and practical experience have taught us that every object has an exact center of mass—a single point located so that the object could be perfectly balanced at that point. This center of mass is known as the object's **centroid**. If we locate the weight of an object at its exact center of mass, or centroid, we can see the effect it will have on the moment about a point in a system.

FIGURE 4.7

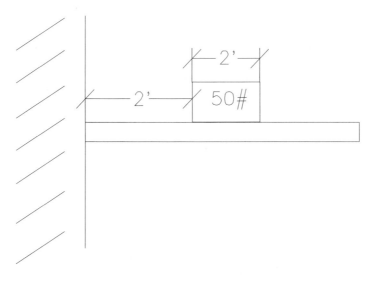

For this portion of the text we will assume that all objects shown have uniformly distributed mass. That is, all parts of the object weigh the same amount per unit volume (have the same density throughout). In such instances, the centroid of the object may be found by the location of its midpoint.

In Figure 4.8 we can see that the centroid of the 50-lb box has been located by finding its midpoint, which is 1 ft from either end. In the left-hand portion of the diagram we see an outline of the box with the location of the centroid being indicated by a 50-lb line of action. We have found the distance of the 50-lb point-force from the fulcrum. This is the distance that we will use to find the moment caused about fulcrum A by the 50-lb force. The moment, then, about fulcrum A is:

$$M_A = -50 \text{ lb} \times 3 \text{ ft}$$
$$= -150 \text{ ft-lb}$$

Up to this point in our study of moments, we have not considered the lever to have any weight. However, in real life, planks, booms, beams, and posts have weight. Sometimes this weight is quite substantial and must always be considered as a moment-causing force.

Figure 4.9 shows a 100-lb block acting as an external force upon a system. The weight of the plank is 10 lb/lf (10 pounds per linear foot), which means that every 1-ft length of this plank has a weight of 10 lb. To determine the overall weight of the plank, we simply multiply the weight per linear foot by the length of the plank in feet:

$$F_{PLANK} = 12 \text{ ft} \times 10 \text{ lb/lf} = 120 \text{ lb}$$

By locating the centroids of both the plank and the block, also shown in Figure 4.9, we can generate the point–force diagram illustrated in Figure 4.10. Finally, we can complete the solution to this problem using a moments chart as shown in Table 4.2.

To better understand the concept of moments as a function of the distance at which a force is applied from a fulcrum, try this simple exercise.

First, place a 12-in. ruler on a small can so that the ruler is centered and balanced on the can. Press lightly on one end of the ruler and watch what happens.

Next, reset the ruler to its original position. Now, press down on the center of the ruler. What happened this time? What does this tell you about moments caused by forces applied at the fulcrum of a system?

MOMENTS 69

FIGURE 4.8

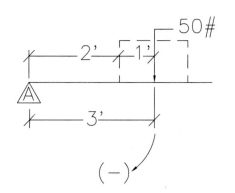
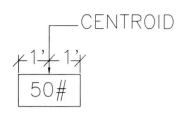

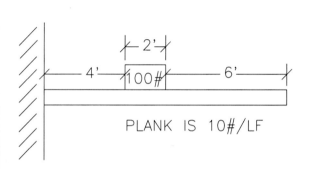
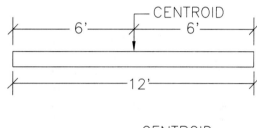
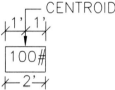

FIGURE 4.9

FIGURE 4.10

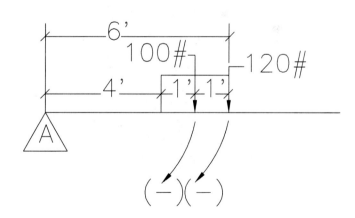

TABLE 4.2

Force	F (lb)	d (ft)	M_A (ft-lb)
1	−100	5	−500
2	−120	6	−720

$\Sigma M_A = -1220$ ft-lb

Stabilizing rotational forces may be accomplished in a number of ways. The most common is by the placement of an external stabilizing force such as a cable or angled support. This is not always necessary, however. The metal used in the design of this traffic signal is light enough to be braced by a series of bolts and a plate located at the junction that prevents rotation at the ground. No external lines or supports are needed. This design is also more flexible and less likely to sustain serious damage during high winds.

4.6 SECOND-ORDER LEVERS

So far all of the lever systems that we have investigated have been first-order levers—levers that have been fastened or supported at one end. In this section we introduce second-order levers. The second-order lever differs from the first-order lever in that the fulcrum may be located anywhere along the line of action so long as it is not located at either end of the lever. Some common uses of the second-order lever are the seesaw and scale balance. The second-order lever is so named because it allows external forces to be placed on either side of the fulcrum. This, in effect, creates two first-order lever scenarios: one to the left of the fulcrum and one to the right.

By examining the lever shown in Figure 4.11 we can see that a portion of the plank extends both to the left and to the right of fulcrum A. Since the

FIGURE 4.11

PLANK IS 5#/LF

OPTION #1   OPTION #2

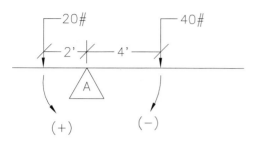 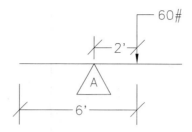

FIGURE 4.12

plank weighs 5 lb/lf we know that it will create some kind of moment about the fulcrum. You have two options when solving for the moments in a second-order lever (Figure 4.12). In option 1 the amount of the total weight of the plank located on each side of the fulcrum is calculated and placed at the centroid of each side. The weight of each side of the plank is found by multiplying the weight per linear foot by the length of the plank segment on each side of the fulcrum. Thus, we have

$$F_L = 4 \text{ ft} \times 5 \text{ lb/lf} = 20 \text{ lb}$$
$$F_R = 8 \text{ ft} \times 5 \text{ lb/lf} = 40 \text{ lb}$$

where F_L is the force on the left-hand side and F_R is the force on the right-hand side. Then, using the point–force diagram shown as option 1 in Figure 4.12, we compute the moments about A due to the forces on both the left and right sides of the fulcrum:

$$M_L = -20 \text{ lb} \times -2\text{ft} = 40 \text{ ft-lb}$$
$$M_R = -40 \text{ lb} \times +4\text{ft} = -160 \text{ ft-lb}$$

where M_L is the moment due to the force on the left-hand side and M_R is the moment due to the force on the right-hand side. Finally, to find the sum of moments about A we simply add the known individual moments:

$$\Sigma M_A = M_L + M_R = 40 \text{ ft-lb} + (-160 \text{ ft-lb}) = -120 \text{ ft-lb}$$

In option 2 the entire weight of the plank is calculated and is then located as a point–force at the centroid of the whole plank (Figure 4.12). Using this procedure we find that the overall weight of the plank is

$$F_{\text{PLANK}} = 12 \text{ ft} \times 5 \text{ lb/lf} = 60 \text{ lb}$$

One application of the second-order lever principle in commercial construction is the use of cantilevers. This reinforced concrete parking garage is designed with cantilevered sections that extend from its main body in the right-hand portion of this photograph. The breaking strength of the concrete cantilever must also be taken into account using the methods that are discussed in Chapter 5.

Then, by locating the 60-lb point–force at the plank centroid located 6 ft from either end of the plank, we can find the distance between the 60-lb point–force and fulcrum A. Finally, we can solve for ΣM_A:

$$\Sigma M_A = -60 \text{ lb} \times 2 \text{ ft} = -120 \text{ ft-lb}$$

Either method will yield the same solution. Notice that in each case the distance from the fulcrum to the point of action needed to be found. This is the distance used to compute the moment about the fulcrum due to that force.

4.7   STABILIZING FORCES

We study the forces acting on various structures so that we may determine how best to stabilize those structures. Our design goal is always to create static equilibrium while maintaining structural integrity with a given system. Since most systems do not automatically fall into static equilibrium upon their initial design, we are left with the responsibility of stabilizing those systems using a variety of methods. The initial method is mathematical in nature. By computing the value of existing forces and moments that act on a system, we can determine the magnitude and direction of the stabilizing force(s) required to place the system in static equilibrium.

FIGURE 4.13

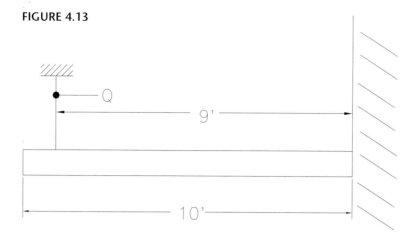

Later attempts to refine the stabilization of the system may include computer-generated models, real-life scale models, and tests conducted on existing structures or a similar structure. In the scope of this course we will study only the mathematical aspect of this intricate process.

Let us begin with a first-order lever scenario. In Figure 4.13 we see a 10-ft long plank weighing 20 lb/lf attached to a wall at its right-hand end. Cable Q, shown near the left end of the plank, will be used to support that end and subsequently to place this system in static equilibrium.

At the outset of any stabilizing force problem we initially ignore the stabilizing force until the latter portion of the problem. The appropriate point–force diagram for this problem is shown in Figure 4.14. Notice that the stabilizing point–force Q is shown as a dashed line in this diagram. This indicates that it will not be used in the computation of the value for ΣM_A. With that in mind,

FIGURE 4.14

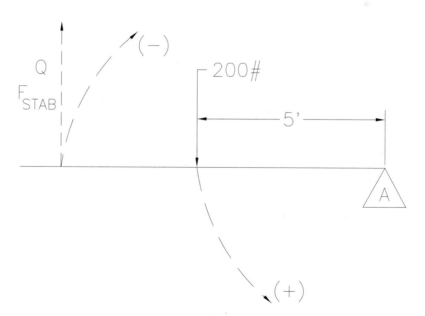

let us solve for ΣM_A due to the known acting force(s) in this system. As shown in Figure 4.14, the only acting force is the weight of the plank itself, which is

$$F_{ACT} = 20 \text{ lb/lf} \times 10 \text{ ft} = 200 \text{ lb}$$

Then the moment about A due to the acting forces only is

$$\begin{aligned} M_A &= F_{ACT}d \\ &= -200 \text{ lb} \times -5 \text{ ft} \\ &= 1000 \text{ ft-lb} \end{aligned}$$

The stabilizing force Q is located 9 ft to the left of the fulcrum. To compute its strength we simply rewrite the basic moments equation to isolate the stabilizing force F_{STAB}:

$$F_{STAB} = \frac{-\Sigma M_A}{d_{A-Q}}$$

Substituting 1000 ft-lb for M_A and -9 ft for d_{A-Q}, we find

$$\begin{aligned} F_{STAB} &= \frac{-1000 \text{ ft-lb}}{-9 \text{ ft}} \\ &= 111.11 \text{ lb} \end{aligned}$$

Therefore, the cable Q must have a minimum tensile strength of 111.11 lb and must pull upward at its present location to stabilize this system.

The concept of system stabilization is similar for second-order lever problems. The main difference is that there are two sides of the lever to work with and, consequently, there may be more than one appropriate location at which a stabilizing force may be applied.

Let us look at the makeshift crane in Figure 4.15. The boom arm weighs 50 lb/lf, and a 100-lb, 6-ft-long box is located 10 ft from the end of a 2-ft-wide support post. We will use as our stabilizing force a set of 25-lb counterweights. Our task will be to determine how many counterweights we will need and at what point along the boom they should be placed.

Figure 4.16 reveals the two options that can be used to set up the solution for this problem. In option 1 the weight of the entire boom is located at its centroid. Since the boom weighs 50 lb/lf and is 28 ft long, it weighs a total of 1400 lb. The fulcrum for this problem is located at the center of the support post. All distances to the existing acting and reacting forces will be measured from this point, shown as point A in Figure 4.16. Finally, the 100-lb block is replaced by a point–force located 14 ft to the right of the fulcrum.

Next, we will use Table 4.3 to find the ΣM_A. If we use option 2 to find ΣM_A, then the weight of the boom will be divided between the two sides of the fulcrum. We see the point–force diagram for this option on the right-hand side of Figure 4.16. The left side of the boom is 11 ft at 50 lb/lf and the right side is 17 ft at 50 lb/lf. These two sides weigh 550 lb and 850 lb, respectively. The weight and location of the 100-lb block are the same as those shown in option 1. The centroid of the left side of the boom lies 5.5 ft to the left of the fulcrum, while the centroid of the right side is located 8.5 ft to the right of the fulcrum.

MOMENTS △ **75**

FIGURE 4.15

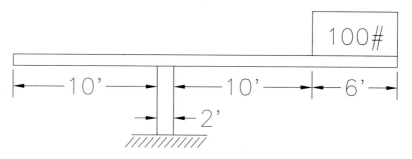

BOOM ARM IS 50#/LF

OPTION #1   OPTION #2

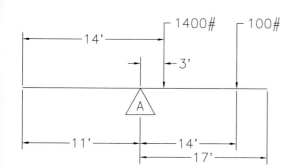   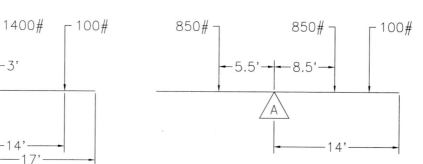

FIGURE 4.16

△ △ **TABLE 4.3**

Force	F (lb)	d (ft)	M_A (ft-lb)
1	−1400	3	−4200
2	−100	14	−1400
			$\Sigma M_A = -5600$ ft-lb

△ △ **TABLE 4.4**

Force	F (lb)	d (ft)	M_A (ft-lb)
1	−850	8.5	−7225
2	−100	14	−1400
3	−550	−5.5	3025
			$\Sigma M_A = -5600$ ft-lb

Using the above values we can generate Table 4.4 to find ΣM_A. Notice again that the ΣM_A is the same regardless of the option used. From this point forward we will use option 2 for the solution of all second-order lever problems in this text.

Now, we must determine how many counterweights we need and where to place them. The counterweights must be utilized as a downward force and as such will need to be placed on the left side of the boom. This will cause a counterclockwise moment in opposition to the existing clockwise moment of 5600 ft-lb that we just found. Since we have 10 ft to work with we can try a variety of locations. In this example we investigate two locations. You may try other weight and location combinations on your own. We want to be sure to use whole counterweights, so we need to place them at a distance from the fulcrum that will divide into the moment of 5600 ft-lb by a whole number. Let us first try 5.6 ft to the left of the fulcrum. At that point we will need

$$F_{STAB} = \frac{5600 \text{ ft-lb}}{-5.6 \text{ ft}}$$

$$= -1000 \text{ lb}$$

The stabilizing force at this point must be 1000 lb downward. By dividing a 25-lb/counterweight into 1000 lb we find that at this point we will need 40 counterweights to balance this crane. If we try again at 7 ft to the left of the fulcrum we determine the required stabilizing force there to be

$$F_{STAB} = \frac{5600 \text{ ft-lb}}{-7 \text{ ft}}$$

$$= -800 \text{ lb}$$

It would require only 800 lb or 32 counterweights at this point to stabilize this crane.

These are only two of the many possible solutions to this problem. Try to find other solutions on your own.

4.8 NORMAL FORCES

Thus far in Chapter 4 we have looked at the effect that certain forces have as they act on a lever within a system. All of the forces that we have looked at have been perpendicular to the LOI. Not all forces, however, are so cooperative in the real world. Many forces are applied at non-right angles to surfaces or lines of interest. For this reason we need to examine the concept of normal forces. A normal force is defined as the component of a force that is perpendicular to a given surface.

Recall from Chapter 2 that the components of any angled force can be calculated if the magnitude and direction of the force are known. Once the components have been found we can see that there is an F_x perpendicular to the y axis and an F_y perpendicular to the x axis. Therefore, a force that lies along one axis is perpendicular to the other axis:

$$F_x \perp y \text{ axis} \quad \text{and} \quad F_y \perp x \text{ axis}$$

We can say then that any vertical force is perpendicular to a horizontal surface, and conversely, any horizontal force is perpendicular to a vertical surface.

To illustrate this concept more clearly, let us examine Figure 4.17. In the top portion of the figure there are three angled forces acting on a horizontal surface. Since the horizontal surface lies along the x axis, we will need to find the vertical component of each force. Likewise, in the bottom portion of the diagram the surface is vertical and the horizontal component of each force is needed.

These two scenarios are useful, but what if the surface lies neither on the vertical nor horizontal axis? Do we need to perform further and more complex calculations to find a force normal to that surface? No, we don't! Look at Figure 4.18. Notice that in solving for the normal forces applied to the horizontal surface we used

$$F_y = F_T \sin \alpha$$

where F_T is the magnitude of the angled force, and α is the acute angle made with the surface. Using this concept we can generate a simple formula to solve for the normal force F_N applied to a surface at any angle:

FIGURE 4.17

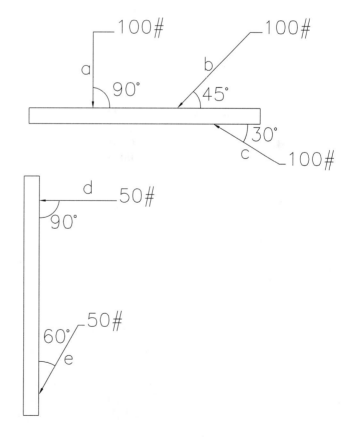

FIGURE 4.18

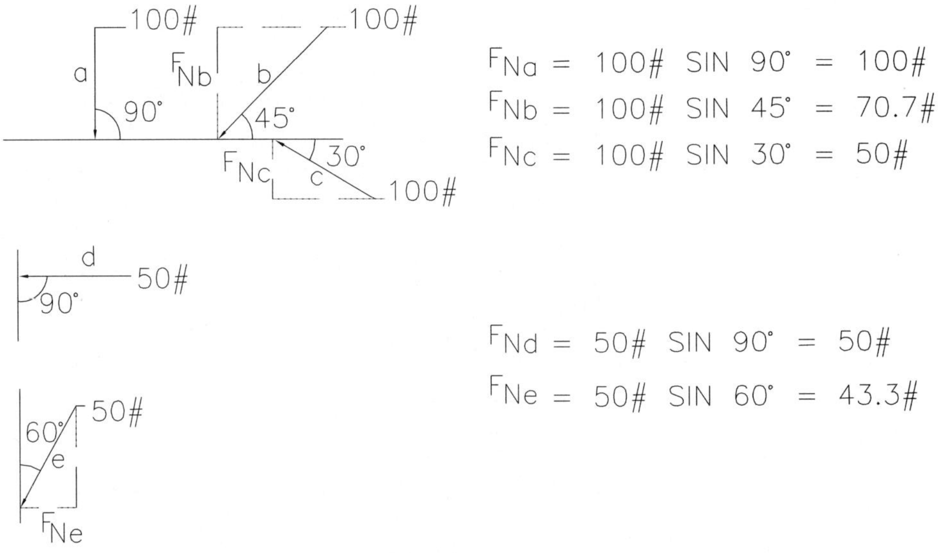
100#
100#
a
FNb
b
90°
45°
FNc
30°
c
100#
FNa = 100# SIN 90° = 100#
FNb = 100# SIN 45° = 70.7#
FNc = 100# SIN 30° = 50#
d
50#
90°
FNd = 50# SIN 90° = 50#
FNe = 50# SIN 60° = 43.3#
50#
60°
e
FNe

$$F_N = F_T \sin \psi$$

where F_T is the magnitude of the angled force, and ψ is the acute angle made with the surface.

Now we can look at a practical application of normal forces in a first-order lever scenario. A support plank, shown on the left side of Figure 4.19, weighs 2 lb/lf and must support an angled load of 10 lb applied at a 30° angle to the plank. What will the sum of the moments be about the fixed point A? First, we label our forces. If we do so from the fulcrum outward, then F_1 becomes 20 lb (because the plank is 2 lb/lf and is 10 ft long), and F_2 is the 10-lb angled force.

As you know by now, only forces perpendicular to a surface will create a moment, so we need to consider the normal forces created by each of these acting forces. To do so, we will slightly alter the moments formula previously given so that only the component perpendicular to the surface will be considered. If we then substitute F_N for F in the moments formula, we get

$$M_A = F_N d$$

and substituting for F_N the formula found in this section we now have

$$M_A = F_T \sin \psi d$$

Now, the chart we use to find ΣM_A will look like Table 4.5. Refer to the lower portion of Figure 4.19 for the point–force diagram and computation of F_N for F_2.

If you look carefully at the previous two examples you will see that the forces that are perpendicular to the given surface need no computation to find F_N. This is true because $\sin 90° = 1$; therefore, any force perpendicular to a surface is already normal to that surface.

FIGURE 4.19

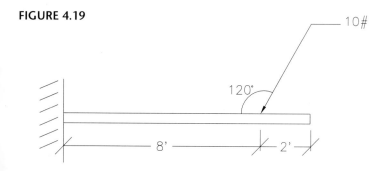

Plank is 2#/ LF

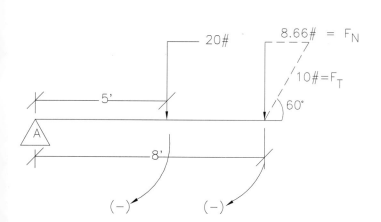

$F_N = F_T \text{ SIN } \psi$

$F_N = 10\# \text{ SIN } 60°$

$F_N = 8.66\#$

△ △ TABLE 4.5	Force	$F_N = F_T \sin \psi$ (lb)	d (ft)	M_A (ft-lb)
	1	−20	5	−100
	2	−8.67	8	−69.36

$$\Sigma M_A = -169.36 \text{ ft-lb}$$

4.9 ANGLED STABILIZING FORCES

The final force that we look at in this chapter regarding moments is a stabilizing force that is applied at an angle to the LOI. All forces have the same effect on a system having moments regardless of whether they are acting or stabilizing forces. In either case, only the component of the force that is normal to the LOI will affect the value of the moments in that system. To study how this works let us refer to Figure 4.20. In this figure a 6-ft steel plank is used to support an 800-lb stationary load while a cable (Q) is run at a 30° angle to the plank and is used to support the load. Our task is to compute the minimum tensile strength of cable Q.

FIGURE 4.20

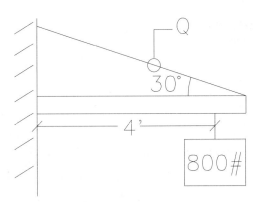

Q
30°
4'
800#
PLANK IS 6' LONG & WEIGHS 40#

First, we need to generate a point-force diagram for all of the acting forces in this system while initially ignoring cable Q. This will yield the point–force diagram shown on the left side of Figure 4.21. Table 4.6 was used to find ΣM_A due to the acting forces in this system. F_1 is the 40-lb weight of the plank located at its centroid, and F_2 is the 800-lb weight located 4 ft from the wall.

Recall from Section 4.7 that the formula for finding a stabilizing force at a known location from the fulcrum is

$$F_{STAB} = \frac{-\Sigma M_A}{d_{A-Q}}$$

This formula allows us to find the value and direction of the force that must act normal to a given LOI at a known point in order to stabilize that system. Since we are dealing with a normal force at this point, but our stabilizing

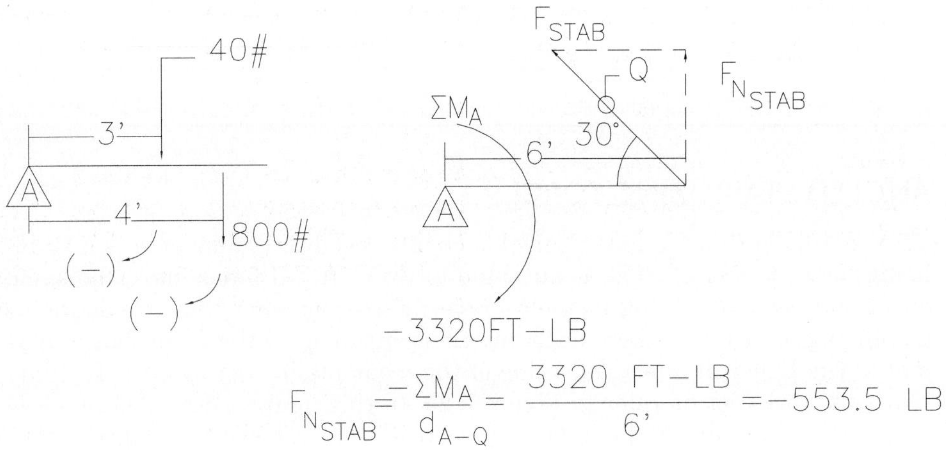

40#
3'
A
4'
800#
(−)
(−)
F STAB
Q
F N STAB
ΣM A
30°
6'
A
−3320FT−LB
F N STAB = −ΣM A / d A−Q = 3320 FT−LB / 6' =−553.5 LB

FIGURE 4.21

TABLE 4.6

Force	F (lb)	d (ft)	M_A (ft-lb)
1	−40	3	−120
2	−800	4	−3200

$$\Sigma M_A = -3320 \text{ ft-lb}$$

cable is obviously not normal to the plank, let us clarify this point by rewriting the above equation to read

$$F_{\text{NSTAB}} = \frac{-\Sigma M_A}{d_{A-Q}}$$

We have incorporated the letter N into the subscript of the stabilizing force to indicate that only the normal component of the stabilizing force is being found at this point. Using this formula we arrive at a value of 553.3 lb for the normal component of this stabilizing force (computation shown in Figure 4.21 at bottom right). This tells us that cable Q must pull at 30° with enough force to generate a normal component of 553.3 lb upward at its current location from fulcrum A.

To determine the actual tension on cable Q along its line of action we use the same formula that we would to determine the normal component of an angled force. This formula is found in Section 4.8.

$$F_N = F_T \sin \psi$$

When we substitute F_{NSTAB} for F_N and rewrite the formula to isolate F_T as the stabilizing force (since F_{STAB} is now acting along the hypotenuse of the force triangle), we arrive at

$$F_{\text{STAB}} = \frac{F_{\text{NSTAB}}}{\sin \psi}$$

The computation necessary to find the actual tension on cable Q is shown along with the stabilizing force triangle in Figure 4.22. The minimum tensile strength for cable Q then is 1106.6 lb.

FIGURE 4.22

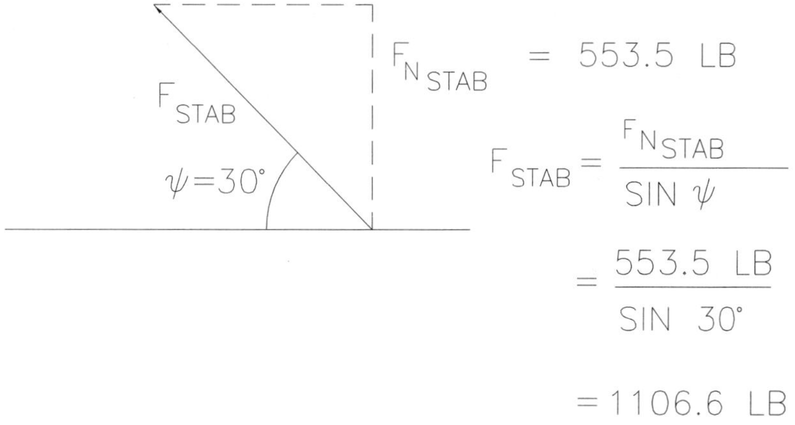

$F_{N_{\text{STAB}}} = 553.5$ LB

$$F_{\text{STAB}} = \frac{F_{N_{\text{STAB}}}}{\sin \psi}$$

$$= \frac{553.5 \text{ LB}}{\sin 30°}$$

$$= 1106.6 \text{ LB}$$

4.10 SUMMARY

As we have seen, a moment is actually a type of rotational work or energy that occurs when a force is applied on an object at any distance from a point on that object that is fixed. We call that fixed point the fulcrum. The units for moments are the same as those for work and energy (ft-lb, kip-ft, g-cm, etc.). Since work is a result of a force applied to a system, and since moments are a type of work, we can conclude that the conditions that hold true for forces and work in a system that is in static equilibrium must also hold true for moments. This conclusion leads to the following:

Condition 3: For a system to be in static equilibrium, the sum of all the moments acting within the system must be zero.

Formula: $\Sigma M = 0$.

We have also discovered that, like forces and work, moments may be computed individually and then added together. The use of a chart such as the ones shown in this chapter makes this task much easier and allows you to organize your work. Moments follow the convention that clockwise (cw) = (−) and counterclockwise (ccw) = (+).

Finally, we have seen the various effects of normal and angled forces on first- and second-order levers. Let us now examine some problems involving both first- and second-order levers in the following sample exercises.

Sample Exercise 4.1

A merchant wants to hang a 100-lb sign in front of his store on a post that extends out from the store wall to a distance of 6 ft. The sign hangs on the post as shown in Figure 4.23. If the wall mounting bracket has a torsional strength of 250 ft-lb and the post weighs 50 lb, how strong will cable Q need to be to support the sign?

As shown in Figure 4.24, three moments are currently acting on this system: (1) the 50-lb post at 3 ft (centroid), (2) the 100-lb sign at 5 ft (centroid), and (3) the 250 ft-lb strength of the mounting bracket. All of these moments are accounted for in Table 4.7. We find that the sum of the moments acting on this system is −400 ft-lb. Cable Q will be needed to counteract this condition.

Looking at Figure 4.25 we find that cable Q is located 6 ft to the right of the fulcrum. We must first find the normal stabilizing force at 6 ft. Then, using

$$F_{STAB} = \frac{F_{NSTAB}}{\sin \psi}$$

we can compute the actual tensile strength of the cable. Solving for the normal stabilizing force we find its value to be

$$F_{NSTAB} = \frac{-\Sigma M_A}{d_{A-Q}}$$

FIGURE 4.23

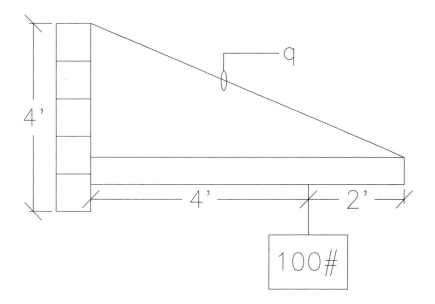

FIGURE 4.24

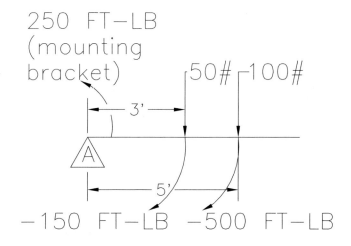

△ △ **TABLE 4.7**

Force	F (lb)	d (ft)	M_A (ft-lb)
1	−50	3	−150
2	−100	5	−500
3			+250
			$\Sigma M_A = -400$ ft-lb

FIGURE 4.25

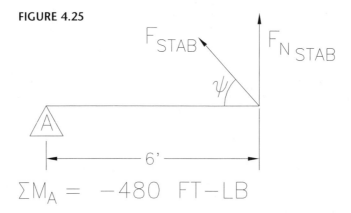

So

$$F_{NSTAB} = \frac{-(-400 \text{ ft-lb})}{6 \text{ ft}}$$

Therefore, the normal stabilizing force = 66.67 lb.
  Next, we need to find the value of angle ψ. We do this by using

$$\psi = \tan^{-1}\left|\frac{4 \text{ ft}}{6 \text{ ft}}\right|$$

Therefore, $\psi = 33.69°$.

  Finally, we find the tensile strength of cable Q by using

$$F_{STAB} = \frac{66.67 \text{ lb}}{\sin 33.69°}$$

Therefore, the tensile strength of cable Q is 120.19 lb.

Sample Exercise 4.2

Compute the torsional value of deceleration required for the adjustable tension pivot bolt of a seesaw to be used by children in the weight range of 35 lb–125 lb if the seesaw board is 8 ft long.

Note: The purpose for the design value of kinetic friction in this problem is to allow variable deceleration of the apparatus so that the heavier child will not fall too quickly nor have an unfair mechanical advantage over the smaller child.

To solve this problem we move directly to the point–force diagram shown in Figure 4.26. By placing the smallest child at one end and the largest child at the other, we can investigate the worst-case scenario, which will allow us to determine the strength range of the adjustable tension pivot bolt. Table 4.8 yields the sum of the acting forces in this problem.

FIGURE 4.26

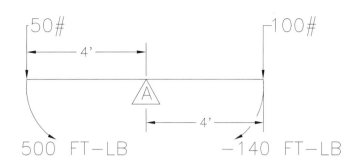

TABLE 4.8	Force	F (lb)	d (ft)	M_A (ft-lb)
	1	−125	−4	500
	2	−35	4	−140
				ΣM_A = 360 ft-lb

Therefore, the torsional limit of the pivot bolt is 360 ft-lb. This condition could be accomplished by the use of a manually adjusted torsion spring mounted at the apparatus support.

Sample Exercise 4.3

A radio tower is 144 ft tall. It rests on a concrete base and is balanced by 3 sets of guy wires. The first set is attached one-third of the way up along the tower, the second is attached one-half of the way up, and the last is attached three-fourths of the way up. All lines are anchored at points 75 ft away from the tower (see Figure 4.27). If each cable has a tensile strength of 10,000 lb and the base anchor has a torsional strength of 50,000 lb, what is the maximum normal load that this tower can resist if the load is applied at its highest point?

Referring to Figure 4.28 we can examine the point–force diagram for this problem. The height at which each cable is attached to the tower has been noted. The tower itself is the LOI and the base is the fulcrum. Each cable is attached to the tower at a different angle, ψ_1, ψ_2, or ψ_3. The values for these angles are found using the procedure shown below:

$$\psi_1 = \tan^{-1}\left(\frac{75}{108}\right) = 37.78°$$

$$\psi_2 = \tan^{-1}\left(\frac{75}{72}\right) = 46.17°$$

$$\psi_2 = \tan^{-1}\left(\frac{75}{48}\right) = 57.38°$$

Each normal force can be found by using the formula $F_N = F_T \sin \psi$:

$$F_{N1} = 10{,}000 \text{ lb} \sin 37.78° = 6126.31 \text{ lb}$$
$$F_{N2} = 10{,}000 \text{ lb} \sin 46.17° = 7213.98 \text{ lb}$$
$$F_{N3} = 10{,}000 \text{ lb} \sin 57.38° = 8422.64 \text{ lb}$$

FIGURE 4.27

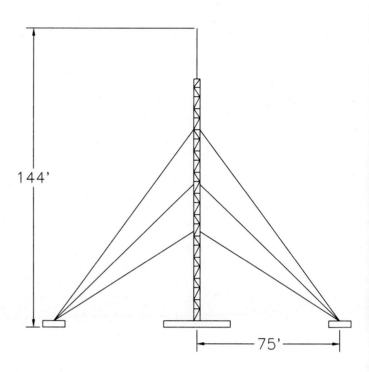

FIGURE 4.28

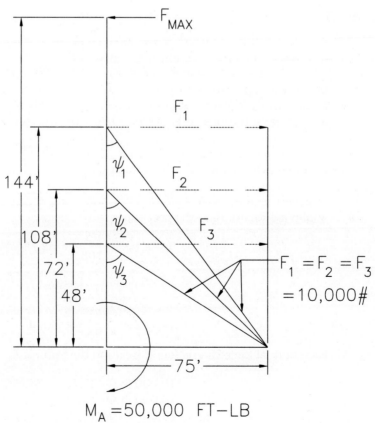

TABLE 4.9	Force	F (lb)	d (ft)	M_A (ft-lb)
	1	6126.31	108	661,641.48
	2	7213.98	72	519,406.56
	3	8422.64	48	404,286.72
	4			50,000
				ΣM_A = 1,635,334.86 ft-lb

In Figure 4.28 the torsional strength of the base anchor has been indicated, since it will also have an effect on the moments applied to this structure. All forces caused by the cable, as well as the moment generated by the base anchor, act to create moments in the same direction. Therefore, all of these moments will be added together as shown in Table 4.9. This value represents the total potential that the tower structure has to resist being bent or rotated about its base. To find the maximum normal load that can be applied to the top of the tower we simply divide the sum of moments about its base by the height of the tower:

$$F_{MAX} = \frac{-\Sigma M_A}{d_{A-FMAX}}$$

Therefore,

$$F_{MAX} = \frac{1,635,334.86 \text{ ft-lb}}{144 \text{ ft}}$$

$$= 11,356.49 \text{ lb}$$

Problems:

In Problems 1–7, solve for M_A or ΣM_A.

1.

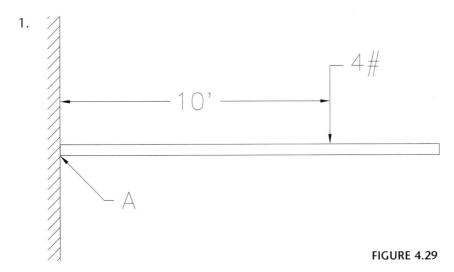

FIGURE 4.29

2.

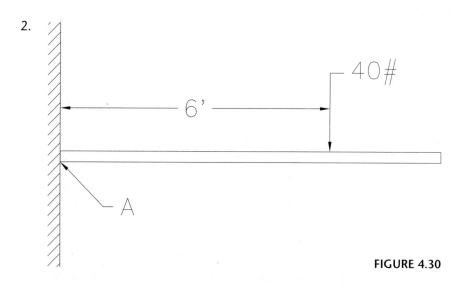

FIGURE 4.30

3.

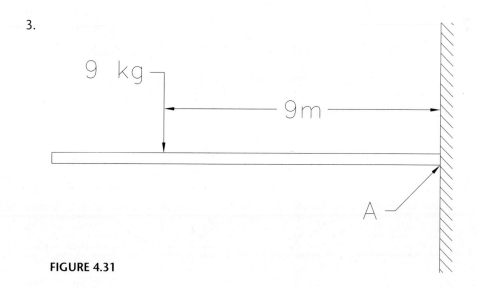

FIGURE 4.31

4.

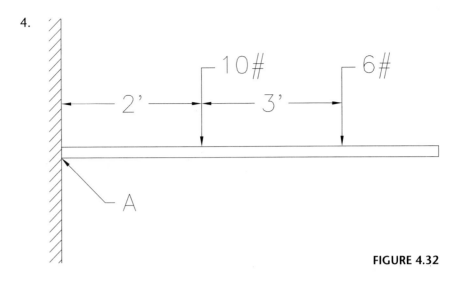

FIGURE 4.32

5.

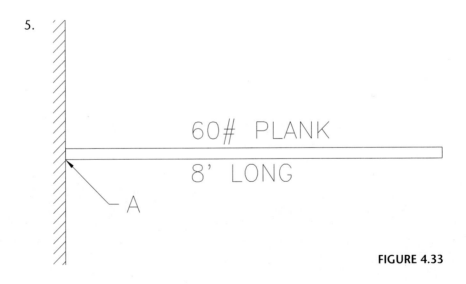

FIGURE 4.33

6.

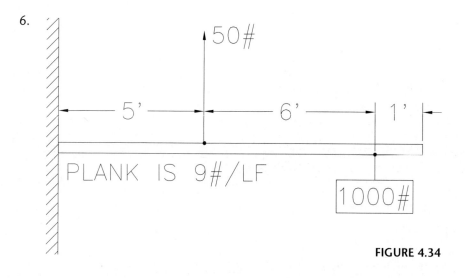

FIGURE 4.34

7.

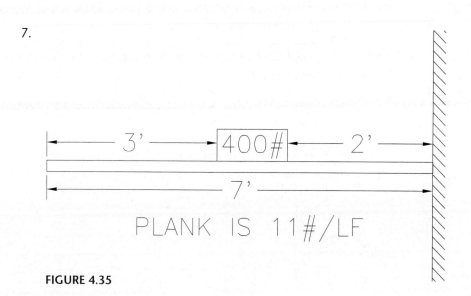

FIGURE 4.35

8. Think of two methods that could be used to stabilize each of the systems shown in Problems 6 and 7.

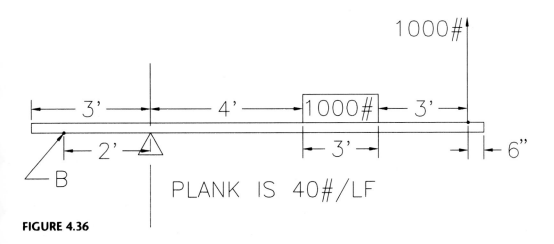

FIGURE 4.36

Problems 9–13 refer to Figure 4.36.

9. Compute the sum of moments about the fulcrum of this system.

10. Explain how a spring with a 500-lb tensile strength can be used to stabilize this system.

11. Explain how a spring with a 700-lb compressive strength can be used to stabilize this system.

12. Show how a cable with a tensile strength of 1000 lb could be used to stabilize this system.

13. Determine the value and direction of a stabilizing force that could be located a point B.

Compute the sum of moments about the fulcrum in Problems 14 and 15.

14.

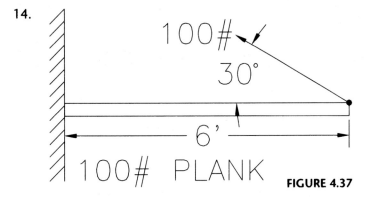

FIGURE 4.37

15.

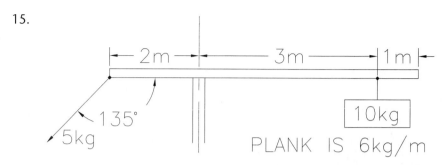

FIGURE 4.38

16. A crane has a 25-ft boom weighing 5000 lb. The maximum load capacity for the crane is 35,000 lb. The boom pulley is located 17 ft from the center of the crane support along the boom. The remaining 8-ft section of the boom extends beyond the support in the opposite direction. How many 2-ton counterweights should be provided with this crane if the counterweight attachment extends 30° downward from horizontal from the end of the 8-ft boom section?

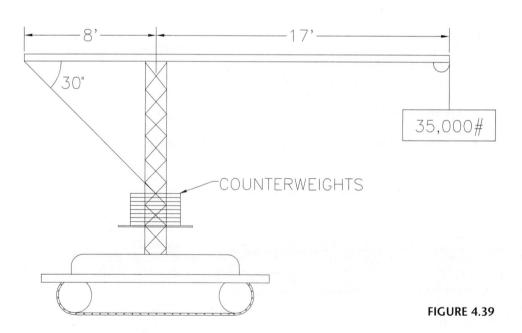

FIGURE 4.39

CHAPTER 5

Beam Forces

5.1 INTRODUCTION

In this chapter we investigate the effects that various forces have on horizontal beams. The following areas of study will be examined:

1. Beam forces and reactions
2. Beam reactions at various fixed points due to moments about a single fixed point
3. Beam reactions in a nonuniformly distributed load scenario
4. Column compression due to
   (a) Internal forces (average)
   (b) External forces

5.2 BEAM FORCES AND REACTIONS DEFINED

We investigate the effects of both internal and external forces acting on a horizontally spanning beam. For this chapter these forces are defined specifically in the following manner:

Internal beam forces—All forces on the beam system due solely to the weight of the beam.

External beam forces—All forces on the beam system due to any and all forces applied from outside the beam.

It is important that these definitions be understood, because internal and external forces may affect a beam system in different ways.

Steel beams like these carry a variety of loads within a structure. The strength of each beam determines how well it will be able to react at the points where it has been attached to other parts of the structure. *Courtesy of MML Construction.*

Newton's third law of motion states that for every action there exists an equal and opposite reaction. For a system in static equilibrium each action and reaction must be exactly equal and opposite with regard to the system in question. For this reason we will study how a beam reacts at fixed points in response to forces applied on it.

A **beam reaction** is a stabilizing force that is a direct response by the beam itself to forces applied on it. Its only purpose is to prevent motion of the beam while it is being stressed. Beam reactions are computed by finding the response of a beam at a fixed point caused by the beam's apparent rotation about another fixed point. Studying the reactions of a beam at various fixed points will help us to determine size, strength, and type of material that should be used for a particular structural element. Beam reaction studies may also be used in the analysis of an existing system to determine if the current structural integrity is sufficient for additional loads applied to the system due to proposed changes in that system.

5.3 COMPUTATION OF BEAM REACTIONS

Since a beam reaction is defined as a stabilizing force, we will treat it as such. Therefore, the basic formula with which we will work is the F_{STAB} formula found in Section 4.7:

$$F_{\text{STAB}} = \frac{-\Sigma M_{\text{A}}}{d_{\text{A-Q}}}$$

FIGURE 5.1

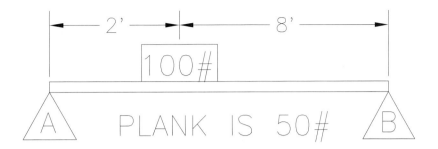

To explain how beam reactions work, let us consider the system shown in Figure 5.1. If we look at the formula shown above and consider the definition of a beam reaction we will realize immediately that we need to find ΣM_A and/or ΣM_B, since a moment is required to solve the problem. Let us first consider the reaction required by support B due to the moments generated about point A. We write this expression in the following manner:

$$R_B \Rightarrow \Sigma M_A$$

This expression is read, "The reaction at B due to the sum of the moments about A."

Using the information shown in the point-force diagram in Figure 5.2, we can solve for ΣM_A. Table 5.1 yields this result. Likewise, Figure 5.3 and Table 5.2 show the solution for ΣM_B. We will save this information for later in the problem.

FIGURE 5.2

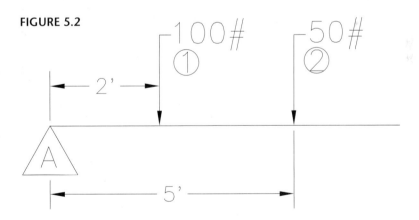

△ △ **TABLE 5.1**

Force	F (lb)	d (ft)	M_A (ft-lb)
1	−100	2	−200
2	−50	5	−250

$\Sigma M_A = -450$ ft-lb

FIGURE 5.3

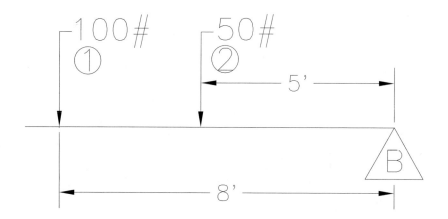

TABLE 5.2

Force	F (lb)	d (ft)	M_A (ft-lb)
1	−100	−8	800
2	−50	−5	250
			$\Sigma M_B = 1050$ ft-lb

To solve for $R_B \Rightarrow \Sigma M_A$, we simply treat the R_B as F_{STAB}, thereby yielding the formula for the solution of this reaction:

$$R_B \Rightarrow \Sigma M_A = \frac{-\Sigma M_A}{d_{A-B}}$$

The distance d between points A and B is 10 ft, but in this solution we must also be careful to consider the *direction* in which this measurement is taken. In the above formula the distance is indicated by d_{A-B}, which means the distance from A to B. The distance must be measured in the correct direction or the direction of the solution for the reaction will be in error.

Using the formula given above for the reaction at B, where $\Sigma M_A = -450$ ft-lb and $d_{A-B} = +10$ ft, we find that:

$$R_B \Rightarrow \Sigma M_A = \frac{-(-450 \text{ ft-lb})}{10 \text{ ft}}$$

$$= +45 \text{ lb}$$

This tells us that the beam will be subjected to a downward force of 45 lb at support B due to the rotation caused about support A and, therefore, must exert an *upward* force of 45 lb to remain in static equilibrium.

We can compute the reaction at support A due to the moments about B ($R_A \Rightarrow \Sigma M_B$) using the same procedure:

$$R_A \Rightarrow \Sigma M_B = \frac{-\Sigma M_B}{d_{B-A}}$$

△ △ **TABLE 5.3**

Acting forces	Beam reactions (lb)
100-lb box	105
50-lb beam	45
150 lb = ΣF_{ACT}	150 lb = ΣF_{REACT}

Notice that this time the distance is measured from B to A (d_{B-A}). This will make the distance from B to A -10 ft because it is now measured from right to left. This yields

$$R_A \Rightarrow \Sigma M_B = \frac{-1050 \text{ ft-lb}}{-10 \text{ ft}}$$

$$= +105 \text{ lb}$$

This tells us that the beam will be subjected to a downward force of 105 lb at support A due to the rotation caused about support B and, therefore, must exert an *upward* force of 105 lb to remain in static equilibrium.

To see if our answers make sense, we refer back to Figure 5.1. Notice that the 100-lb weight is located closer to support A. This indicates that support A will be required to carry a greater load than support B, which is farther away from the weight. According to our answers, this is shown to be true.

Also, since the sum of all forces in a system in static equilibrium must add up to zero, the sum of our upward reactions must equal the sum of our downward actions. This is illustrated in Table 5.3.

Beam Reactions due to Nonuniformly Distributed Masses

Thus far, all of the moment and beam reaction problems that we have studied have involved uniformly distributed masses. These masses have the same weight per lineal unit, unit of area, or unit of volume. A beam weighing 60 lb/lf is an example of a **uniformly distributed mass** because every one-foot section weighs 60 lb. There is no variation in weight from one end to the other.

In **nonuniformly distributed loads**, however, this is not true. One section may weigh more than another section, and that variation may occur at specific points or throughout the entire length of the object. An example of a beam having a **nonuniform mass** is shown in Figure 5.4. The beam is made up of three sections of uniformly distributed masses, but is itself a nonuniformly distributed mass. Notice that the left-most portion weighs much more than the right-most portion, while the center portion is heavier still.

FIGURE 5.4

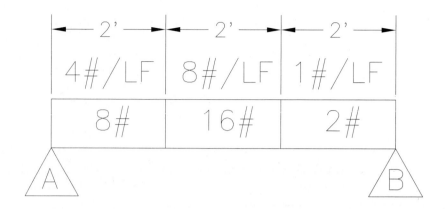

How do we compute moments and reactions within such a scenario? First, we generate a point–force diagram treating each section as an individual internal force. Let us illustrate this procedure by first finding $R_B \Rightarrow \Sigma M_A$.

The point–force diagram for this procedure is shown in Figure 5.5. The chart for computing ΣM_A is shown as Table 5.4. Now, to find $R_B \Rightarrow \Sigma M_A$ we simply use the appropriate formula:

$$R_B \Rightarrow \Sigma M_A = \frac{-(-66 \text{ ft-lb})}{6 \text{ ft}}$$

$$= +11 \text{ lb}$$

FIGURE 5.5

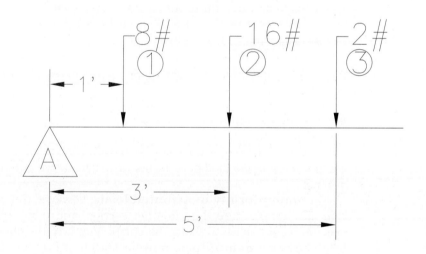

TABLE 5.4

Force	F (lb)	d (ft)	M_A (ft-lb)
1	−8	1	−8
2	−16	3	−48
3	−2	5	−10

$\Sigma M_A = -66$ ft-lb

Looking ahead to the next support we would expect the reaction at A to be greater than the one at B because there is more weight on the left side of the plank. Referring to Figure 5.6 and Table 5.5 we see the point–force diagram and calculations associated with the computation of ΣM_B. For $R_A \Rightarrow \Sigma M_B$ we find

$$R_A \Rightarrow \Sigma M_B = \frac{-(90 \text{ ft-lb})}{-6 \text{ ft-lb}}$$

$$= +15 \text{ lb}$$

Notice that the reaction at support A is just a bit more than that at support B. This is because there is more weight on support A than on support B.

Note: The reaction values calculated in the previous examples do not reflect the actual compression values on the supports at the given points; rather, they represent the reaction by the beam itself at those points. These kinds of reactions, known as **column compressions**, are discussed in the next section of this chapter.

FIGURE 5.6

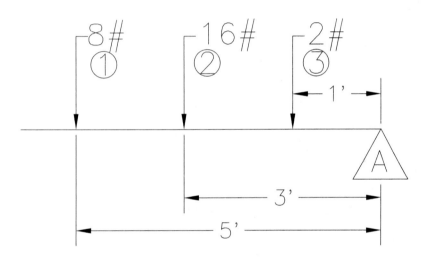

TABLE 5.5	FORCE	F (lb)	d (ft)	M_B (ft-lb)
	1	−8	−5	40
	2	−16	−3	48
	3	−2	−1	2
				ΣM_B = 90 ft-lb

Nonuniformly distributed loads are common in commercial and residential construction alike. The house roof shown in this pkotograph is an example of such a load. This house reveals a heavier load at the peak of this oddly shaped roof. The main horizontal supports (joists or beams) are concealed within the structure and are subjected to varying loads throughout their lengths. It would be possible to construct block diagrams for these roofs to illustrate the varying loads that they carry.

Nonuniformly distributed forces need not always be located on the line of interest. They may be stacked to represent areas of high density or nonuniformly distributed external loads. To further study this phenomenon let us refer to Figure 5.7. The **block diagram** in the top portion of Figure 5.7 is the same diagram as the one found in Figure 5.4! In Figure 5.7 the diagram uses stacked blocks of equal size to represent forces that are indicated by weight per linear foot. In a **block diagram** each linear foot of an object is repre-

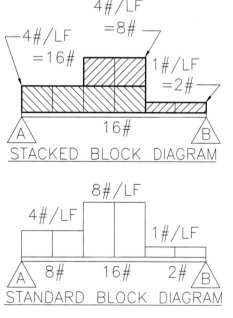

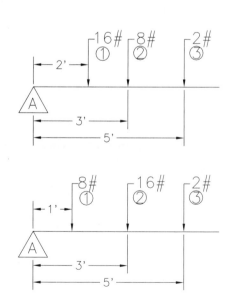

FIGURE 5.7

sented by a single block. Blocks having a higher weight per linear foot value may be either taller or stacked to represent the greater density of that portion of the object. Both Figures 5.4 and 5.7 are equally correct representations of the same force scenario. Choice of application is left solely to the discretion of the individual performing the calculations.

5.4 COMPUTATION OF COLUMN COMPRESSIONS DUE TO INTERNAL FORCES

We continue our study of structures by investigating the effects that various forces applied to a system have on the vertical supporting members called **columns**. Figure 5.8 shows a generic example of a beam supported at either end by columns. If the beam load is uniformly distributed, then we can easily compute the compression on both columns A and B. We do this by determining the approximate average value for the load that each column will carry. To compute this in a uniformly distributed load scenario we divide the beam into two parts, locating the division exactly halfway between the two supports. In doing so we discover that each support will carry exactly one-half of the total load of the beam, so that if the beam weight is F then the compression on each column due to the weight of the beam is $F/2$.

If the beam support columns are not located at equal distances from the center of the beam or if there are more than two supports, then it may be necessary to perform further calculations to determine the compression on each of the columns. For instance, if the location of column B shown in Figure 5.9 is on the centroid of the beam, then it would appear that each column would carry one-third of the total load. Right? Let's see.

FIGURE 5.8

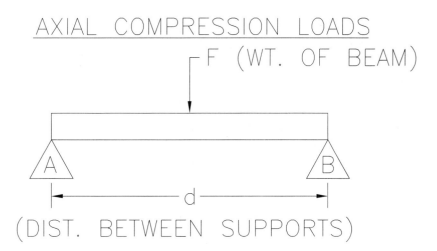

FIGURE 5.9

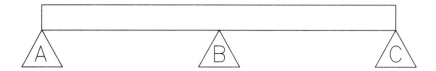

If we divide the beam at all points located exactly halfway between each pair of supports while giving the beam an overall length of 12 ft and a total weight of 240 lb, we will arrive at the result shown in Figure 5.10. The resulting weights of each section are shown in the diagram. Each weight corresponds to the average force that each segment will apply to the column on which it rests. Therefore, the corresponding column compressions are as follows:

The average compression on column A: $C_A = 60$ lb
The average compression on column B: $C_B = 120$ lb
The average compression on column C: $C_C = 60$ lb

Figure 5.11 illustrates the method used to divide a beam for maximum load distribution resting on unevenly spaced supports. Assuming a beam weight of 10 lb/lf, we would arrive at the following column compressions:

$$C_A = 3 \text{ ft} \times 10 \text{ lb/lf} = 30 \text{ lb}$$
$$C_B = 5 \text{ ft} \times 10 \text{ lb/lf} = 50 \text{ lb}$$
$$C_C = 5 \text{ ft} \times 10 \text{ lb/lf} = 50 \text{ lb}$$

Notice that if we add all of the column compressions together they will be equal to the total weight of the beam:

$$C_A + C_B + C_C = 30 \text{ lb} + 50 \text{ lb} + 50 \text{ lb} = 130 \text{ lb}$$
$$\text{Beam weight} = 13 \text{ ft} \times 10 \text{ lb/lf} = 130 \text{ lb}$$

FIGURE 5.10

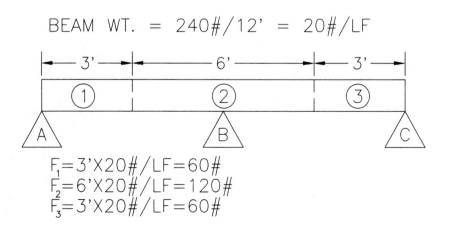

FIGURE 5.11

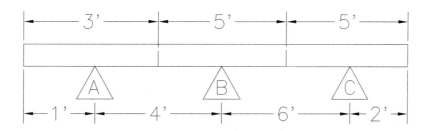

These two angled steel posts help carry the load transmitted by the steel I-beam above. Analyses of the reactions and compressions occurring at these points due to the loads applied from above are important steps in the design of a structural support such as this one.

5.5 COMPUTATION OF COLUMN COMPRESSIONS DUE TO EXTERNAL FORCES

External forces applied to a beam, such as weights, cables, and springs, affect the compression on supports differently than the weight of the beam itself does. One aspect of external loads that serves to simplify calculations concerning their contribution to the compression on the beam supports is that external forces compress only the two columns between which they are located.

The beam in Figure 5.12 has two external loads on it. The 100-lb load is located exactly halfway between supports A and B, whereas the 200-lb weight is located closer to support C than to support B. We would, therefore, expect

FIGURE 5.12

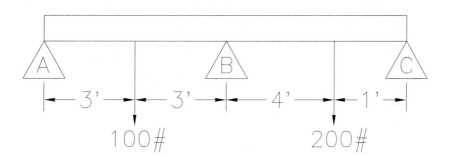

FIGURE 5.13

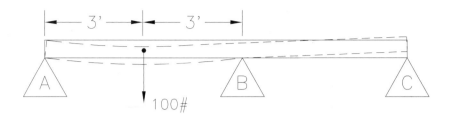

the loads on A and B due to the 100-lb force to be equal, while the loads on B and C due to the 200-lb force would be unequal. Let us examine this problem.

To properly assess the total compression on the supports caused by the external forces we must first investigate the effects of each force individually on the supports. The 100-lb force, for example, will affect only supports A and B. This is true because of the deflection that takes place between supports A and B, as shown in Figure 5.13. Notice the dashed lines representing the effect that the moments about A and B have on the beam itself. Only supports A and B are compressed.

To compute the actual value of compression on each of the supports we can create a chart that looks like Table 5.6. The distance that goes in the numerator of the fraction is measured from the force to the *opposite* support, not the support for which the compression is being computed. The distance that appears in the denominator of the fraction is the distance between the two support points. Both distances are absolute values and are, therefore, always treated as positive numbers.

In algebraic notation the formula for the two calculations shown in the previous chart is

$$C_A = F \frac{|d_{B-F}|}{|d_{A-B}|}$$

where

C_A = compression on support A due to force F
F = external force located between A and B
d_{A-F} = distance from force F to support B
d_{A-B} = distance from support A to support B

TABLE 5.6

Point	Force (lb)	×	Distance to opposite support (ft)	÷	Distance between points (ft)	=	C (lb)
A	100		3		6		50
B	100		3		6		50

$\Sigma C_{A\&B}$ = 100 (lb)

FIGURE 5.14

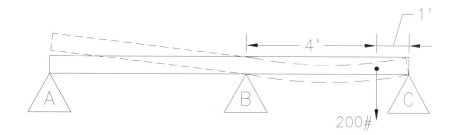

Armed with this formula let us return to the scenario in Figure 5.11 and continue to investigate the compressions due to the 200-lb force. As we now know, this force will compress only supports B and C as shown in Figure 5.14. Since it is located much closer to support B we would expect the greater compression to occur there. Also, remember that both the 100-lb force and the 200-lb force cause compression on support B. We will investigate this condition shortly.

Table 5.7 provides the results of the compressions on B and C due to the 200-lb force. The sum of the compressions is equal to the value of the individual forces. This is a quick check to see if any blatant errors have been made during calculation, but it is not a substitute for a thorough check of the work performed. Be sure to perform this check for each problem you solve, remembering that it is possible to have the sum of your compressions add up to the total force value while having incorrect solutions for the individual compressions.

The problem at hand is not yet completed. We have found compressions at all three points, but we have not yet totaled the results. This is done as follows:

$$\begin{aligned}
\Sigma C_A &= 50 \text{ lb} \\
\Sigma C_B = 50 \text{ lb} + 100 \text{ lb} &= 150 \text{ lb} \\
\Sigma C_C &= \underline{100 \text{ lb}} \\
\Sigma C &= 300 \text{ lb}
\end{aligned}$$

The results show us the load distributions of the external forces on the three supports. As you might expect, the load on support B is the greatest because it is being compressed by both forces. The sum of the compressions is shown also as a preliminary check of the work.

TABLE 5.7

Point	Force (lb)	×	Distance to opposite support (ft)	÷	Distance between points (ft)	=	C (lb)
B	200		1		5		40
C	200		4		5		160
							$\Sigma C_{B\&C}$ = 200 lb

FIGURE 5.15

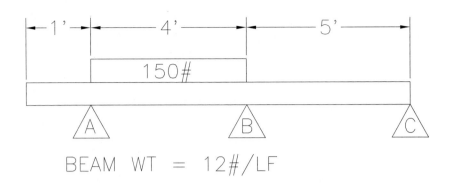

Now, let us investigate a system that has both internal and external forces acting on it so that we may study how the combination of these forces affects the compression on its supports. Referring to Figure 5.15, we see a beam resting on three supports. The beam has a uniformly distributed internal mass of 12 lb/lf and an external force of 150 lb resting on it. To solve a problem such as this we generate a point–force diagram and solve for compressions due to internal and external forces separately. Figure 5.16 shows the point–force diagram with all critical distances labeled. Using this overall dimensioned diagram we can solve for both the internal and external compressive forces. First, let's look at the internal forces.

The outermost numbers represent the segments into which the beam is to be divided in order to compute the internal force compressions. These lengths are 3, 4.5, and 2.5 ft. The resulting internal compressions are, then

$$C_A = 3 \text{ ft} \times 12 \text{ lb/lf} = 36 \text{ lb}$$
$$C_B = 4.5 \text{ ft} \times 12 \text{ lb/lf} = 54 \text{ lb}$$
$$C_C = 2.5 \text{ ft} \times 12 \text{ lb/lf} = 30 \text{ lb}$$

Since the 150-lb external force covers the entire distance from A to B, its centroid will be exactly halfway between those two supports. Each support will, then, carry one-half of the 150-lb load, or 75 lb each. The external compressions on A and B are shown in Table 5.8.

To find the total compression on each support we add all of the individual compressions:

FIGURE 5.16

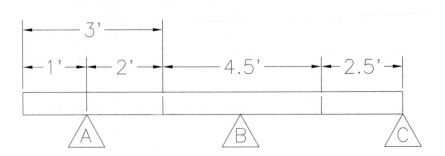

△ △ **TABLE 5.8**

POINT	FORCE (LB)	×	DISTANCE TO OPPOSITE SUPPORT (FT)	÷	DISTANCE BETWEEN POINTS (FT)	=	C (LB)
A	150		2		4		75
B	150		2		4		75

$$\Sigma C_A = 36 \text{ lb} + 75 \text{ lb} = 111 \text{ lb}$$
$$\Sigma C_B = 54 \text{ lb} + 75 \text{ lb} = 129 \text{ lb}$$
$$\Sigma C_C \phantom{= 54 \text{ lb} + 75 \text{ lb}} = 30 \text{ lb}$$
$$\Sigma C = 270 \text{ lb}$$

To verify the total load we add the weight of the beam to the weight of the external force:

$$10 \text{ ft} \times 12 \text{ lb/lf (internal)} + 150 \text{ lb (external)} = 270 \text{ lb}$$

Use the above procedure when computing column compressions due to internal and external forces combined. By solving for each compression individually you reduce your risk of error in calculation.

5.6 SUMMARY

We have now seen the fourth and final condition that must be met for a system to be in static equilibrium:

Condition 4: For a system to be in static equilibrium, the sum of all the reactions within the system must be equal to the sum of all the actions upon the system.

Formula: ΣActions = ΣReactions

This condition clearly implies that the total sum of all actions and reactions in a system in static equilibrium must be zero. This confirms Newton's third law of motion, which states that for every action there must be an equal and opposite reaction. In this case the result of all of the reactions is a statically stable system.

Let us now look at a sample problem that involves all of the topics covered in this chapter.

The metal sliding boards shown here are supported at various locations to counteract the variable loads that they have been designed to carry. Notice that at certain points the ground itself acts as the support for the load. Since these sliding boards have been built into the side of a hill, their supports have been placed perpendicular to that hill. How do you think that the concept of normal forces (discussed in Chapter 4) would affect the design of these boards?

Sample Exercise 5.1

For the system shown in Figure 5.17 solve for the following:

(a) $R_A \Rightarrow \Sigma M_B$
(b) $R_B \Rightarrow \Sigma M_A$
(c) $\Sigma C_A, \Sigma C_B, \Sigma C_C$

(a) To solve for $R_A \Rightarrow \Sigma M_B$, we first need to find ΣM_B. To do this we will use the chart shown on the following page while referring to the point–force diagram shown in Figure 5.18. The centroid of the 14-ft beam is at 7 ft from either end, making it 3 ft from fulcrum B, while the centroid of the 100-lb weight is located 4 ft to the right of B. Using the diagram we can generate Table 5.9.

Next, we insert the above value into the beam reaction equation from Section 5.3.

$$R_A \Rightarrow \Sigma M_B = \frac{-(-1450 \text{ ft-lb})}{-4 \text{ ft}}$$

$$= -362.5 \text{ lb}$$

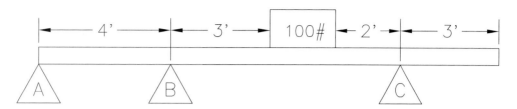

FIGURE 5.17

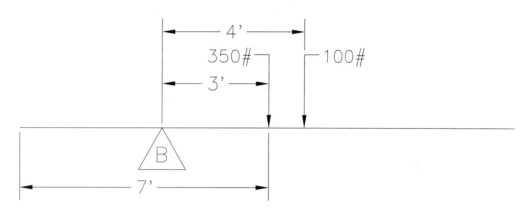

FIGURE 5.18

△ △ **TABLE 5.9**

Force	F (lb)	d (ft)	M_B (ft-lb)
1	−350	3	−1050
2	−100	4	−400
			$\Sigma M_B = -1450$ ft-lb

(b) The solution for $R_B \Rightarrow \Sigma M_A$ is accomplished in exactly the same manner as for that of the previous problem. The point–force diagram for finding ΣM_A is shown in Figure 5.19 and the subsequent solutions that follow are shown in Table 5.10.

$$R_B \Rightarrow \Sigma M_A = \frac{-(-3250 \text{ ft-lb})}{4 \text{ ft}}$$

$$= -812.5 \text{ lb}$$

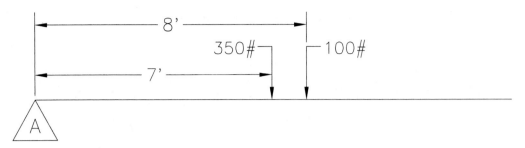

FIGURE 5.19

△ △ **TABLE 5.10**

Force	F (lb)	d (ft)	M_A (ft-lb)
1	−350	7	−2450
2	−100	8	−800
			$\Sigma M_A = -3250$ ft-lb

(c) Finally, we need to solve for the compressions on each of the supports. As noted earlier, we should solve for internal and external compressions independently, then find their sum on each support. Let us begin by looking at the compressions due to the internal forces only. Referring to Figure 5.20, we can see the linear distribution of each segment of the beam after it has been divided at the midpoints between each support.

To compute the load that each support carries due to the weight of the beam we simply compute the weight of each segment. The beam weighs 25 lb/lf, as shown in Figure 5.20. To find the compression of the beam on each support we do the following:

$$C_A = 2 \text{ ft} \times 25 \text{ lb/lf} = 50 \text{ lb}$$
$$C_B = 5.5 \text{ ft} \times 25 \text{ lb/lf} = 137.5 \text{ lb}$$
$$C_C = 6.5 \text{ ft} \times 25 \text{ lb/lf} = 162.5 \text{ lb}$$

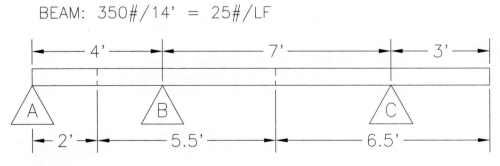

FIGURE 5.20

Since the single external load is located between supports B and C, it will affect only those two supports in compression. Figure 5.21 shows the point–force diagram for this external force. The computations for the distribution of this external force are shown in Table 5.11. Remember, support A is not affected by the 100-lb external force. To find the total compression on each support we add all of the individual compressions:

$$\Sigma C_A = 50.00 \text{ lb}$$
$$\Sigma C_B = 137.5 \text{ lb} + 42.86 \text{ lb} = 180.36 \text{ lb}$$
$$\Sigma C_C = 162.5 \text{ lb} + 57.14 \text{ lb} = \underline{219.64 \text{ lb}}$$
$$\Sigma C = 450.00 \text{ lb}$$

To verify the total load let's add the weight of the beam to the weight of the 100-lb external force:

$$350 \text{ lb (internal)} + 100 \text{ lb (external)} = 450 \text{ lb}$$

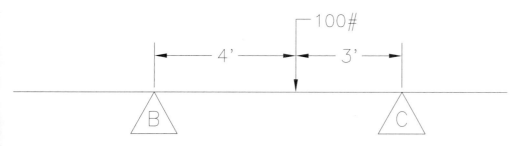

FIGURE 5.21

△ △ **TABLE 5.11**

Point	Force (lb)	×	Distance to opposite support (ft)	÷	Distance between points (ft)	=	C (lb)
B	100		3		7		42.86
C	100		4		7		57.14

$$\Sigma C_{B-C} = 100 \text{ LB}$$

FIGURE 5.22

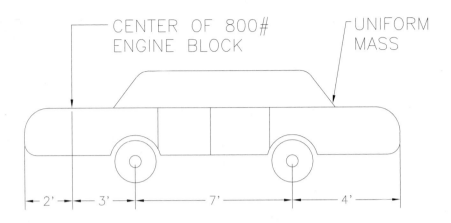

Sample Exercise 5.2

The car shown in Figure 5.22 weighs 1.9 tons. Determine the

(a) reaction at the front axle
(b) reaction at the rear axle
(c) compression on each axle treating the weight of the car as an external force

(a) Figure 5.23 shows the point–force diagram needed to determine ΣM_R, which is the sum of moments about the rear axle. Table 5.12 displays the result of these moments. To find R_F we divide ΣM_R by the distance from R to M:

$$R_F \Rightarrow \Sigma M_R = \frac{-32{,}000 \text{ ft-lb}}{-7 \text{ ft}}$$

$$= 4571 \text{ lb (rounded to the nearest pound)}$$

FIGURE 5.23

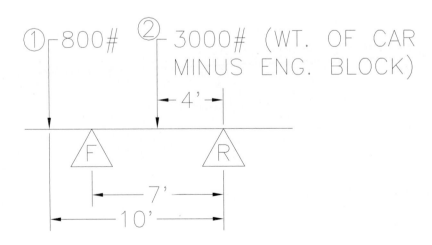

TABLE 5.12

Force	F (lb)	d (ft)	M_A (ft-lb)
1	−800	−10	8,000
2	−3000	−4	24,000
			ΣM_B = 32,000 ft-lb

(b) We repeat this procedure to find $R_R \Rightarrow \Sigma M_F$ (Figure 5.24, Table 5.13): Then, to find R_R:

$$R_R \Rightarrow \Sigma M_F = \frac{6600 \text{ ft-lb}}{7 \text{ ft}}$$

$$= 943 \text{ lb (rounded to the nearest pound)}$$

(c) Finally, to find the compression on each axle, we treat all loads as external forces. This yields the point–force diagram shown in Figure 5.25. Notice that the 800-lb engine block rests solely on the front axle. This value must be added to the compression due to the weight of the car on that axle. To find the compression on each axle due to the weight of the car we use Table 5.14. Therefore,

$$C_F = 1714 \text{ lb} + 800 \text{ lb} = 2514 \text{ lb}$$
$$C_R = \underline{\phantom{1714 \text{ lb} + 800 \text{ lb}} = 1286 \text{ lb}}$$
$$\Sigma C = 3000 \text{ lb}$$

FIGURE 5.24

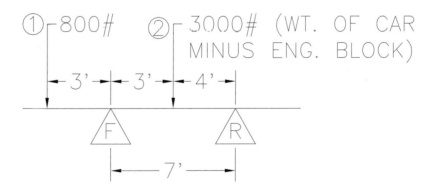

TABLE 5.13

Force	F (lb)	d (ft)	M_A (ft-lb)
1	−800	−3	2400
2	−3000	3	−9000
			ΣM_B = −6600 ft-lb

FIGURE 5.25

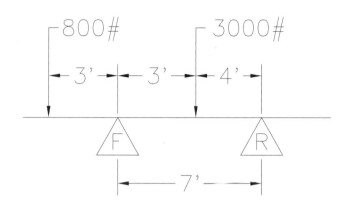

△ △ **TABLE 5.14**

Point	Force (lb)	×	Distance to opposite support (ft)	÷	Distance between points (ft)	=	C (lb)
F	3000		4		7		1714
R	3000		3		7		1286

ΣC = 3000 lb

Sample Exercise 5.3

A reinforced concrete landing having a mean density of 160 lb/ft³, shown in Figure 5.26, is designed to carry a uniformly distributed load of 120,000 lb. How strong does the landing need to be at point B to carry this load?

First, we need to determine the weight of each section of the landing, since these loads will contribute to the moment about the fixed point A. To do this we simply multiply the volume of each section, or block of concrete, by its density of 160 lb/ft³.

$$\text{Left block } (F_1) = 4 \text{ ft} \times 10 \text{ ft} \times 2 \text{ ft} \times 160 \text{ lb/ft}^3 = 12{,}800 \text{ lb}$$
$$\text{Right block } (F_3) = 8 \text{ ft} \times 10 \text{ ft} \times 1.5 \text{ ft} \times 160 \text{ lb/ft}^3 = 19{,}200 \text{ lb}$$

Next, we can draw the point–force diagram for this system as shown in Figure 5.27, placing each force at its centroid. The 120,000-lb design load is placed in the center of the whole platform since it is being treated as a uniformly distributed mass. Table 5.15 shows the resulting sum of moments about A due to these forces.

FIGURE 5.26

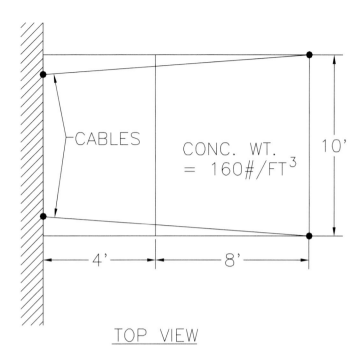

FIGURE 5.27

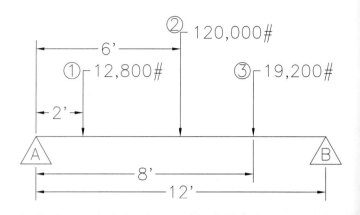

	TABLE 5.15	Force	F (lb)	d (ft)	M_A (ft-lb)
		1	−12,800	2	−25,600
		2	−120,000	6	−720,000
		3	−19,200	8	−158,600
					$\Sigma M_A = -904{,}200$ ft-lb

To find the reaction at point B we simply divide $-\Sigma M_A$ by the distance from A to B.

$$R_B \Rightarrow \Sigma M_A = \frac{904{,}200 \text{ ft-lb}}{12 \text{ ft}}$$

$$= 75{,}350 \text{ lb}$$

But, notice in Figure 5.26 that there are two cables supporting the far end of the platform. This means that each reaction point will be subjected to only one-half the total load to that end.

$$\therefore R_B = 75{,}350 \text{ lb} \div 2 \text{ cables} = 37{,}675 \text{ lb}$$

Problems:

1. In Figure 5.28 find (a) ΣM_A and (b) $R_B \Rightarrow \Sigma M_A$.

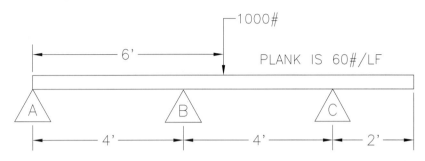

FIGURE 5.28

2. Find the reaction at each support in Figure 5.29 due to the individual moments about all other supports. You will have six total answers—two reactions for each support.

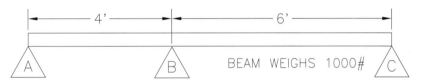

FIGURE 5.29

3. Compute $R_A \Rightarrow \Sigma M_B$.

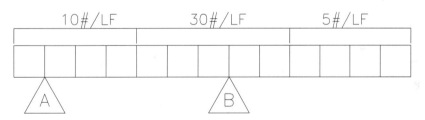

FIGURE 5.30

4. Compute $R_B \Rightarrow \Sigma M_A$ if the 50-lb weight is centered on the plank.

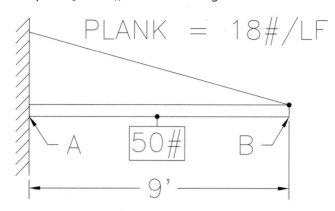

FIGURE 5.31

5. In Figure 5.32, compute (a) the compression on each support due to the beam weight only, (b) the compression on each support due to the external forces only, and (c) the sum of compressions on each support.

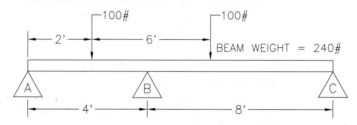

FIGURE 5.32

CHAPTER 6

Steel Beam Design

6.1 INTRODUCTION

In this chapter we study the uses of formulas as they apply to steel structures. The specific areas of study will be

1. Reading steel beam design charts
2. Computing sectional structural loads for
   (a) Existing structures
   (b) New structures
3. Steel beam sizing calculations
4. Steel beam selection based on
   (a) Size criteria
   (b) Load criteria

6.2 STEEL TERMINOLOGY

Steel is an alloy of iron and carbon that made its way into the building construction industry around the turn of the 20th century. Before the advent of steel, iron was the most popular structural metal in use. Its strength and durability made iron ideal for such jobs as bridges, rails, decorative railings, and structural framework. One of the main drawbacks to iron, though, was its rapid tendency to **oxidize,** or rust. It was found that by adding just the right amount of carbon to molten iron, an alloy could be formed that was highly resistant to oxidation. Today, steel is used widely throughout the construction industry. Its structural uses range from steel wall studs and nails in residential settings to I-beams, plates, bolts, and rivets in commercial settings.

Steel can also be modified for many other uses. One example is **stainless steel,** which is an alloy made of standard steel and the addition of chromium

Years of ground settling and thermal shock have caused this retaining wall to crack in a telltale pattern along its mortar lines. This wall is exposed to direct sunlight for most of the day and has little protection from the elements in bad weather. These extreme temperature conditions have aided in the decomposion of the mortar on this wall. The results are obvious.

and nickel. In fact, the single largest use of nickel is in the production of stainless steel. This particular alloy of steel is highly resistant to oxidation and other forms of corrosion and is also highly resistant to **thermal shock**. Thermal shock is the effect that a sudden change in temperature has on a particular material. You may have seen plastic garbage cans or other receptacles that have been left outdoors for long periods of time. After a while they begin to warp and crack. This is the result of thermal shock. These items have expanded and contracted with the wide variation in temperature to which they have been exposed. This will eventually cause material **fatigue** and **failure,** which we discuss later in this chapter.

Since nickel does not readily expand or contract under thermal variations, or changes in temperature, it lends a stabilizing property to stainless steel that helps to prevent **deformation** due to thermal shock. For this reason, stainless steel is used on the exterior of many building structures as decoration, wall plates, and framing elements. Additionally, stainless steel is used in the transfer and storage of many extremely hot or cold substances because it is less likely to react chemically with these substances or become damaged in the process.

Cracks such as the one shown in this concrete slab can be aided in development by exposure to extreme weather conditions.

Some general variations on steel are:

High-carbon steel (carbide)—Steel made with a higher concentration of carbon. This form of steel takes on a blue-black or black color. It is less dense than standard steel, and has a much higher compressive strength.

Heat-tempered steel—This steel is formed by quickly cooling a heated piece of standard steel. The resulting metal is more rigid, but tends to be more brittle than standard steel.

6.3 PROPERTIES OF METALS

Metals are different from all other chemical elements and compounds. They are uniquely distinguishable by three inherent characteristics:

Malleability—The ability of a substance to be struck and reshaped by a hammer. A malleable substance will hold the new shape consistently as it is being formed. Metal's ability to be forged is an example of this property.

Ductility—The ability of a substance to be drawn out, extruded, or folded into a new shape. Metal will characteristically hold the new shape. Ductwork and aluminum soda cans are a couple of examples of the ductility of metal.

Conductivity—The ability of a substance to transfer heat and/or electricity. Iron magnets, copper and aluminum wire, and stainless steel cookware are all examples of this property of metals.

One of the three basic properties of all metals is conductivity. Copper and aluminum are particularly good conductors of electricity. For this reason these metals are widely used for electrical power lines, telephone wire, and electric power cables. Wrapping these cables in rubber or electrical tape protects the metal wire from oxidation, and protects us from electrocution!

Conductivity may also be displayed by a material's ability to transfer heat or magnetism. This iron skillet readily transfers heat from a stove to the food placed inside it. Iron is also one of only three metalic elements that is inherently magnetic. The other two magnetic metals are nickel and cobalt.

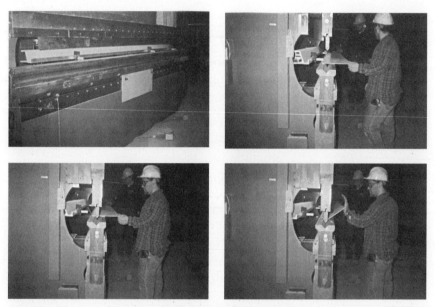

The machine shown in these photographs is used to bend plates of metal into right angles. This illustrates the metallic property of ductility as well as the material property of permanent set: (top left) The bending machine empty, (top right) a metal plate is positioned over the groove, (bottom left) pressure begins to be applied to the plate, (bottom right) the plate is bent into its new shape. *Courtesy of Peirce Welding and Fabrication.*

All metals have each of these three properties to some degree. In addition, most metals, in their pure state, have a shiny luster when polished. Even sodium, which is soft at room temperature, and mercury, which is liquid at room temperature, exhibit these characteristics.

6.4 ARCHITECTURAL STEEL: PROPERTIES AND STRESS ANALYSIS

For the purpose of this course we will be investigating the properties of A36 steel, a grade of steel commonly used in architecture. We will also be investigating the various properties of beams under stress. These properties apply to wood and concrete, which we study in later chapters, as well as steel.

FIGURE 6.1

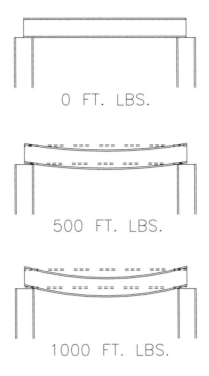

Any substance, whether under external stress or not, experiences **deformation.** Deformation is the change that occurs to the shape and size of an object due to the forces acting upon it. Deformation is directly proportional to the load applied to an object. In the case of a steel beam spanning two supports, as shown in Figure 6.1, the beam bends more as the weight applied to it increases. This bending is known as **deflection.** Deflection is defined as the vertical deviation of a beam over a horizontal span. In Figure 6.2 we see the value of the distance the beam has bent as Δy. The unbraced length of the span is called x. The deflection, then, of the beam shown in Figure 6.2 is

$$\delta = \frac{\Delta y}{x}$$

FIGURE 6.2

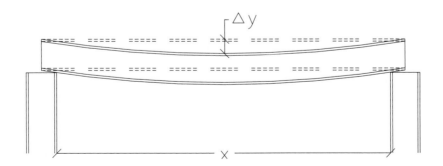

where

δ = deflection value of the beam
Δy = vertical deviation of the beam
x = unbraced length of the beam

For a steel beam spanning two supports to meet industry standards, it must not bend more than $\frac{1}{360}$ of its total unbraced span under a given **live load.** A live load is the load due to additional external forces after a structure has been completed. Some examples of live loads are people, furniture, mobile office equipment, and cars. We will be investigating the effects of live loads on structures from here on in this text. We will assume that all of the permanent loads, or **dead loads,** are already accounted for.

When a beam bends due to the forces applied to it from above, it experiences **flexure.** The fibers on top of the beam are in compression, while the fibers on the bottom of the beam are in tension, as illustrated in Figure 6.3. Flexure will occur at a uniform rate for a certain degree of deflection. This uniform rate of deformation under increasing stresses is known as **Hooke's law,** named after the seventeenth century mathematician and physicist, Robert Hooke, who first discovered this property of materials.

Hooke's law, however, holds true only up to a certain point of deflection. Beyond a certain point of bending, permanent deformation will occur. At that point a beam will weaken somewhat and will begin to deform more rapidly as more stress is added. The weakening of the beam due to its deformation is called **fatigue.** The load required to reach this point of permanent deformation is called the **elastic limit** of the material. The point at which the elastic limit is reached, that is, the point beyond which the beam will no longer bend uniformly, is called its **yield point.** Once a beam has been bent to this point, Hooke's law no longer applies.

When a beam becomes deformed due to the stresses applied to it so that it retains the deformed shape after the stress has been removed, we say that it has a **permanent set.** If we continue to apply an ever-increasing load on a beam, it will eventually reach **failure.** That is, it will collapse and/or break

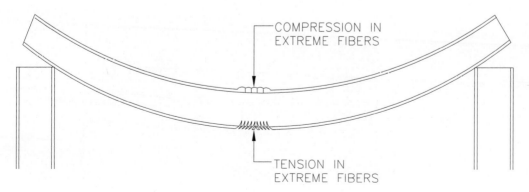

FIGURE 6.3

All of these tools were formed by a process called forging. Each tool has been shaped by some form of impact. For example, the tip of the crowbar was "flanged" by repeated striking, while the body of the wrench was "stamped" or "drop-forged" using an impact machine to stamp out the rough shape of the tool. This process of shaping metals into tools is possible because of the property known as malleability.

apart. When this occurs the **ultimate strength** of the beam has been reached. The ultimate strength of A36 steel is approximately 70,000 psi.

Another property of steel is its **modulus of elasticity**. The modulus of elasticity indicates the stiffness of a material. This property is determined in a laboratory by taking a solid round or square rod of a material and subjecting it to extreme tension. The numerical value for modulus of elasticity E is found using the following formula:

$$E = \frac{Pl}{Ae}$$

where

P = force on the rod in pounds
l = length of the rod in inches
A = cross-sectional area of the rod in square inches
e = elongation of the rod under stress in inches

Rust has formed on the outer surface of these metal plates. Another word for rust is oxidation, a chemical process in which oxygen combines with a metal to form another substance. Ferrous oxide, or iron rust, with its telltale red-orange color, is the most recognizable form of oxidation. *Courtesy of Peirce Welding and Fabrication.*

The modulus of elasticity for A36 steel is 29,000,000 psi, according to the ASTM (American Society for Testing and Materials). As we see later in this chapter, it is necessary to know the modulus of elasticity of a material to determine its size, shape, and strength when bearing a load.

Let us now look at a couple of examples of how the above information can be practically applied. A 14-ft-long I-beam spans two columns and bends $\frac{1}{16}$ in. vertically at its center. Compute the deflection of this beam. Using the formula for deflection from earlier on in this chapter we have

$$\delta = \frac{0.0625 \text{ in.}}{168 \text{ in.}}$$

Notice that the values for Δy and x are both in inches. Both values must always be in the same units to correctly compute deflection. The value for deflection in this problem is

$$\delta = 0.00037$$

The above beam would fall within industry standards because it deflects less than $\frac{1}{360}$ of its span. The decimal value of $\frac{1}{360}$ is 0.00278. Any span having a deflection value less than 0.00278 meets industry standards for live load deflection.

Now, let's look at a condition for which the modulus of elasticity formula can be used. A 10-ft-long steel cable is made up of 500 strands of steel each 0.01 in. in diameter. If it is not allowed to stretch more than 0.03 in., what is the maximum load it can carry?

We begin with the formula for modulus of elasticity shown previously in this chapter. Rewriting the formula to solve for the missing load we find

$$P = \frac{EAe}{l}$$

where

E = 29,000,000 psi (for A36 steel)
A = 500 strands $\times$ $\pi(0.005 \text{ in.})^2$ per strand = 0.03925 in^2
e = 0.03 in.
l = 120 in.

So, then, we have

$$P = \frac{29,000,000 \times 0.03925 \times 0.03}{120}$$

$$= 284.56 \text{ lb}$$

Considering a moderate safety factor, you could hang about 250–280 lb on this cable before it would stretch beyond its allowed limit (safety factors are discussed in greater detail in Chapter 7).

FIGURE 6.4

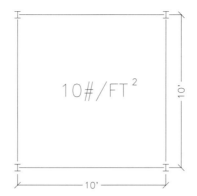

6.5 UNIFORM LOAD DISTRIBUTIONS ON RECTANGULAR SECTIONS

For the purposes of this text we will assume that all loads shown on any section of a structure are uniformly distributed. In this way we can investigate the basic principles of structural design. Figure 6.4 shows a section of a steel framing plan (discussed in Chapter 7) that is 10 ft × 10 ft. The total load on this section is 10 ft × 10 ft × 10 lb/ft^2 = 1000 lb. We will consider this load evenly distributed over each unit of area of the section shown. Therefore, to compute the portion of the load carried by each of the four beams (shown as single lines between the I's) we must compute the area of the section that rests on each beam.

To find the area of each section we divide the full rectangular section into four triangles by scribing the diagonals of the square as illustrated in Figure 6.5. The four triangles intersect at the centroid of the section. Each triangu-

FIGURE 6.5

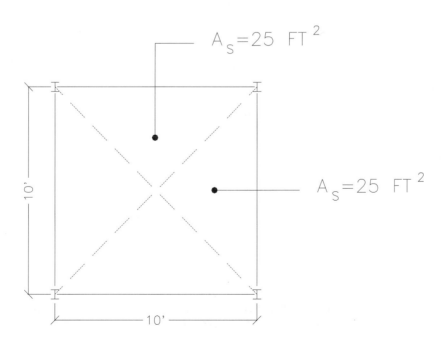

FIGURE 6.6

lar segment represents the portion of the total section that rests on each beam. We then compute the load carried by each beam by computing the area of each triangular segment. In this case, since the section is a square, all four triangles will be equal in area. Therefore, each beam carries one-fourth of the total load, or 250 lb. This seems simple enough because a square has four equal sides. However, let us consider a rectangular section that does not have four equal sides. How do we treat that scenario? Referring to Figure 6.6 we see a building section that is 10 ft × 6 ft. How will the lengths of the sides affect the load that they carry? Using the formula for the area of a triangle and referring to Figure 6.6 we find

$$A = \tfrac{1}{2}bh$$

where

b = length of base
h = height or altitude

Then

$$A_{S1} = \tfrac{1}{2} \times 10 \text{ ft} \times 3 \text{ ft} = 15 \text{ ft}^2$$

and

$$A_{S2} = \tfrac{1}{2} \times 6 \text{ ft} \times 5 \text{ ft} = 15 \text{ ft}^2$$

Notice that, even though the dimensions of the triangles are different, their areas are the same. This holds true for all rectangles regardless of their dimensions. Thus, if the section in Figure 6.7 is uniformly loaded, then each triangular segment would represent the same load.

FIGURE 6.7

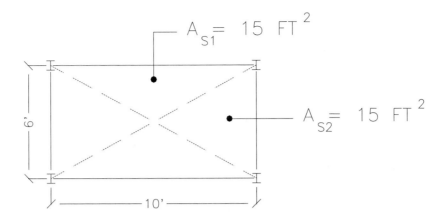

Nonrigid bodies exert a greater force on their supports as they approach their centroids. To simplify this scenario, we eliminate this condition. We will assume at this point that all uniform loads discussed in this text are perfectly rigid.

As you probably already know most buildings are made up of more than one rectangular framed section. An example of a structure made of two equal-sized rectangular sections is shown in Figure 6.8. In this structure the beams labeled A, C, D, E, F, and G each carry one-fourth the total load from their respective sections. Beam B, however, carries a load from both the right- and left-hand sections. To further investigate this condition let us perform a brief analysis on this structure:

$$\text{Weight of left section:} \quad W_{\text{LEFT}} = 8 \text{ ft} \times 8 \text{ ft} \times 10 \text{ lb/ft}^2 = 640 \text{ lb}$$
$$\text{Weight of right section:} \quad W_{\text{LEFT}} = 8 \text{ ft} \times 8 \text{ ft} \times 20 \text{ lb/ft}^2 = 1280 \text{ lb}$$
$$\therefore W_A = W_D = W_E = \tfrac{1}{4}(640 \text{ lb}) = 160 \text{ lb}$$
$$W_C = W_F = W_G = \tfrac{1}{4}(1280 \text{ lb}) = 320 \text{ lb}$$

The weight on B is a result of loads coming from both sections, as shown in Figure 6.9:

$$\therefore W_B = \tfrac{1}{4}(640 \text{ lb}) + \tfrac{1}{4}(1280 \text{ lb})$$
$$= 160 \text{ lb} + 320 \text{ lb} = 480 \text{ lb}$$

FIGURE 6.8

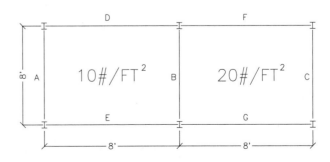

FIGURE 6.9

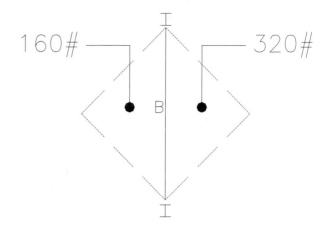

So the load capacity of beam B must be greater than that of the other beams. Notice that beam B is an interior beam. Interior beams are usually required to carry greater loads than exterior beams because they are loaded from both sides. In scenarios involving uniform loads this will always be the case.

Sometimes sides of rectangular sections are subdivided due to the nature of the structure. The entire side of the section will carry one-fourth the total section load, but each beam on that side will not carry the full complement of that load. Let us consider the diagram in Figure 6.10. By now you should have no problem recognizing the procedure to solve for the loads carried by the beams labeled A, D, and E. If we assume a uniform weight of 10 lb/ft^2 then the loads on these three beams would be 200 lb each. The loads on

FIGURE 6.10

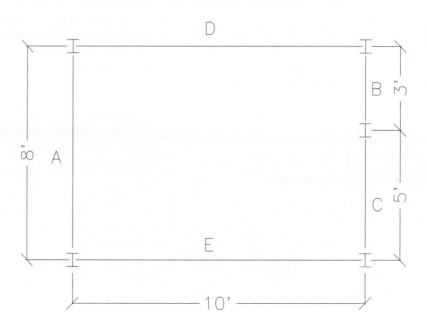

FIGURE 6.11

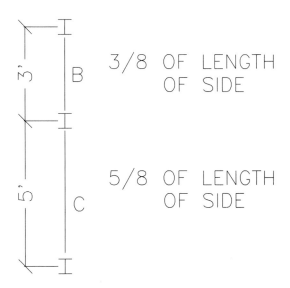

beams B and C, however, would not be found in exactly the same manner. The total load on the side of the section containing these two beams is also 200 lb, but we need to determine how this load is distributed between the two beams.

We do this by first understanding that uniformly distributed loads over an area are also uniformly distributed along the sides of that area. Therefore, the longer a segment on a side of the section the more load it will need to carry. Figure 6.11 illustrates this concept. In the drawing we see the fraction of the total side length accounted for by each beam. Beam B, which is 3 ft long, accounts for $\frac{3}{8}$ of the total length of the side, while beam C, which is 5 ft long, accounts for the other $\frac{5}{8}$ of the total side length. The distribution of loaded segments is proportional to the length of each portion of that segment. So, we can find the load carried by beams B and C in the following manner:

$$W_B = \tfrac{3}{8}(200 \text{ lb}) = 75 \text{ lb}$$
$$W_C = \tfrac{5}{8}(200 \text{ lb}) = 125 \text{ lb}$$

The generic formula for solving for loads on segmented beams is

$$(\text{length of segment} \div \text{length of side}) \times \text{load on side}$$

We can also apply the above principle to structures that have segmented beams that share interior and exterior loads. An example is shown in Figure 6.12. Pay

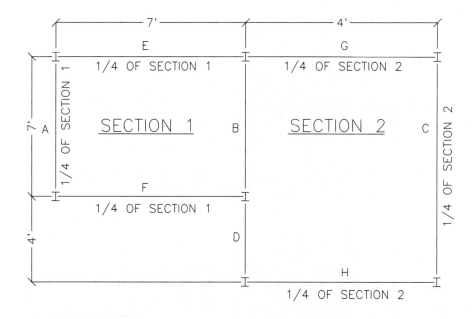

B = 1/4 SECTION 1 + (7/11 × 1/4) SECTION 2
D = 4/11 × 1/4 SECTION 2

FIGURE 6.12

particular attention to the distribution shown for beams B and D. Beam B is an interior beam which shares one-fourth the load on section 1 and a portion of the one-fourth load from section 2. Beam D does not share any of the load from section 1 and carries only a portion of the one-fourth load from section 2. The load distributions are shown in Figure 6.8. Later in this chapter a sample exercise illustrates this situation.

6.6 STEEL BEAM NOMENCLATURE

In *Romeo and Juliet* Shakespeare asked the question, "What's in a name?" In the case of a steel beam, much can be discovered by the name itself. Below is an example of a steel beam name. Let us explore the various parts of its name to see what we can discover.

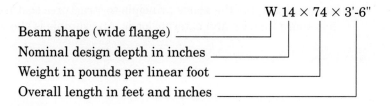

△ △ **TABLE 6.1**

ABBREVIATION	NAME	DESCRIPTION
W	Wide flange	I-beam with wide flanges. Entire shape will fit roughly into a square.
S	Standard	I-beam with standard flanges. Depth of beam is about twice its breadth.
M	Miscellaneous	I-beam having depth-to-breadth ratios between wide flange and standard I-beams.
C	Channel	C-shaped beam.
L	Angle	L-shaped beam. May have equal or unequal legs.
T	Tee	T-shaped beam. Formed by bisecting an I-beam laterally through its web.

Not only do we know the information shown above, but we can also easily find the weight of the beam by multiplying the last two numbers in its name together:

$$74 \text{ lb/lf} \times 3.5 \text{ ft} = 259 \text{ lb}$$

Beams come in many shapes and sizes. The shapes that we will be looking at in this text are listed in Table 6.1.

The cross section of a steel I-beam is a bit more complex than it may appear at first glance. A drawing of an I-beam section with most of its dimensions is shown in Figure 6.13. You may wish to place a reference tab on the page so that you can quickly reference it during future exercises until you become more familiar with the I-beam. In Figure 6.14 you will also find some sketches of the beams listed in Table 6.1. Take some time to familiarize yourself with these dimensions and definitions. The letter representing each dimension on a standard I-beam and corresponding definitions can be found in Table 6.2. We will be using all these definitions and dimensions in Chapters 6 and 7. In fact, some of them are put to use in the next section of this chapter.

FIGURE 6.13

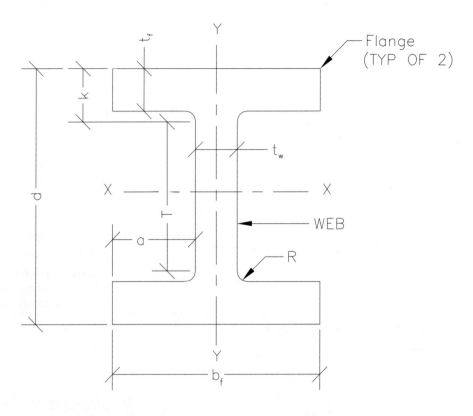

I-BEAM DIMENSIONS

FIGURE 6.14

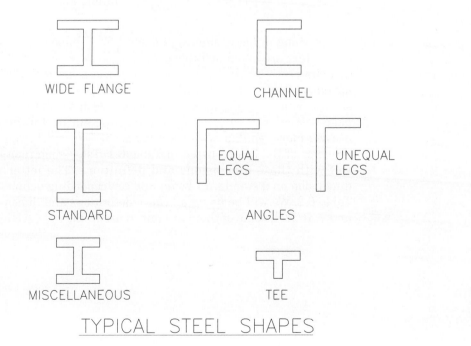

TYPICAL STEEL SHAPES

TABLE 6.2   Dimensions of steel I-beams (from Figure 6.9)

Symbol	Definition
a	Distance from end of flange to face of web
b_f	Breadth of flange
d	Depth of section
k	Distance from face of flange to radius tangency at web
R	Fillet radius between web and flange
T	Length of straight section of web between fillets
t_f	Thickness of flange
t_w	Thickness of web
x-axis	Axis perpendicular to web
y-axis	Axis perpendicular to flange

6.7   CUTTING LENGTHS OF STEEL BEAMS

Steel beams spanning two columns must be cut to fit the space between the columns. For example, if two columns are spaced 20 ft apart on center (O.C.), then the length of the beam spanning those two columns would need to be less than 20 ft. Figure 6.15 illustrates this condition. Notice that the beams must be cut back to account for the thickness of the columns. Notice also that there is a small space between the columns and the ends of the beams. This space must also be accounted for.

All of the dimensions labeled in Figure 6.15 need to be understood to accurately determine the length to which a steel beam spanning two columns must be cut. The following criteria must be determined when performing these calculations:

L—center-to-center distance between columns in inches

tol.—clip tolerance of $\frac{1}{2}$ in. on each end (discussed in greater detail in Chapter 7)

FIGURE 6.15

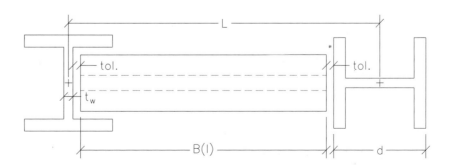

Orientation

Web-to-web—The beam touches the web of both columns.

The thickness of each column web (t_w) must be determined.

Flange-to-flange—The beam touches the flange of both columns.

The depth of section (d) of each column must be determined.

Web-to-flange—The beam touches the web of one column and the flange of the other column.

The t_w value must be determined for the column whose web is touched by the beam and the d value must be determined for the column whose flange is touched by the beam.

Figure 6.15 shows a typical web-to-flange span of a beam between two wide-flange I-beams. The dimension B shown in Figure 6.15 represents the **actual cut length** of the beam in inches.

Let us now examine the three orientations in more detail by looking at an example of each. In Figure 6.16 we see a beam spanning two W14×34 columns in a web-to-web orientation. With this type of orientation we need to account for the web thickness t_w of each of the two columns. The web thickness for a W14×34 ($t_{w(W14\times34)}$) is 0.285 in., as shown in Table A in the back of this text. To determine the actual length to which this beam must be cut we need to subtract the following dimensions from the center-to-center column span (L):

$\frac{1}{2}$-in. clip tolerance against column 1

$\frac{1}{2}$-in. clip tolerance against column 2

One-half the web thickness of column 1 ($\frac{1}{2}t_{w1}$)

One-half the web thickness of column 2 ($\frac{1}{2}t_{w2}$)

Writing this in the format of an algebraic formula we have

$$B = L - \tfrac{1}{2}\text{ in.} - \tfrac{1}{2}\text{ in.} - \tfrac{1}{2}t_{w1} - \tfrac{1}{2}t_{w2}$$

In this instance both columns 1 and 2 are the same size, so we can combine the values of their web thicknesses to arrive at

FIGURE 6.16

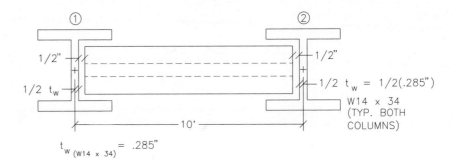

Steel Beam Design 137

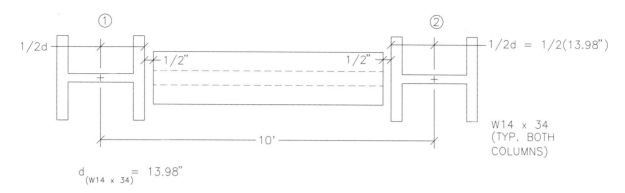

FIGURE 6.17

$$B = L - 1 \text{ in.} - t_w$$

This is the standard formula for computing the cutting length of a beam that is spanning two equally sized columns in web-to-web orientation. If we apply this formula to the current example we have

$$B = 120 \text{ in.} - 1 \text{ in.} - 0.285 \text{ in.}$$
$$= 118.715 \text{ in.}$$

So the beam in this example would have to be cut to a length of 118.715 in. to span the two columns in Figure 6.16.

Let us continue this study by looking at a beam that spans two W14×34 columns that are again 10 ft apart, but this time in a flange-to-flange orientation (Figure 6.17). This time we need to determine the depth of section of the W14×34 columns. From Table A we find this value to be $d_{(W14 \times 34)} = 13.98$ in. Applying the same principle as we did in the previous example we arrive at

$$B = L - 1 \text{ in.} - d$$

for a flange-to-flange span of equally sized columns. Substituting the values we found into the above formula we have:

$$B = 120 \text{ in.} - 1 \text{ in.} - 13.98 \text{ in.}$$
$$= 105.02 \text{ in.}$$

Notice that, even though the beams in Figures 6.16 and 6.17 are spanning columns that are 10 ft apart each, the actual cutting length of the two beams is quite different.

Finally, let's consider a situation like the one shown in Figure 6.18 in which a beam spans two columns 10 ft apart touching the web of one column and the flange of the other. You will notice that for column 1 we need to know the web thickness and for column 2 we will need the depth of section. To determine the actual cut length of this beam we will need to

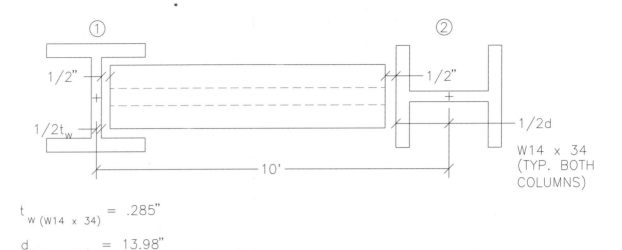

FIGURE 6.18

subtract the standard 1-in. clip tolerance, half the web thickness of column 1 ($\frac{1}{2}t_{w1}$) and half the depth of section of column 2 ($\frac{1}{2}d_2$). Our formula then reads as follows:

$$B = L - 1 \text{ in.} - \tfrac{1}{2}t_{w1} - \tfrac{1}{2}d_2$$

This is the standard formula for computing the cutting length of a beam spanning two columns that is touching the web of the first column and the flange of the second column. If we, then, substitute our known values for t_w and d we have

$$B = 120 \text{ ft} - 1 \text{ in.} - \tfrac{1}{2}(0.285 \text{ in.}) - \tfrac{1}{2}(13.98 \text{ in.})$$
$$= 111.8675 \text{ in.}$$

The standard formulas for all variations of orientation are shown in Table 6.3.

TABLE 6.3

		Beam Cutting Length Formulas		
Symbol	Orientation	Abbreviation	Equal columns	Unequal columns
I-I	Web-to-web	W-W	$B = L - 1 \text{ in.} - t_w$	$B = L - 1 \text{ in.} - \tfrac{1}{2}t_{w1} - \tfrac{1}{2}t_{w2}$
H-H	Flange-to-flange	F-F	$B = L - 1 \text{ in.} - d$	$B = L - 1 \text{ in.} - \tfrac{1}{2}d_1 - \tfrac{1}{2}d_2$
I-H	Web-to-flange	W-F	$B = L - 1 \text{ in.} - \tfrac{1}{2}t_w - \tfrac{1}{2}d$	$B = L - 1 \text{ in.} - \tfrac{1}{2}t_{w1} - \tfrac{1}{2}d_2$

6.8 SIZING STEEL I-BEAMS USING STANDARD LIVE-LOAD DEFLECTION

The standard formula used for computing the deflection of a beam is

$$D = \frac{5Wl^3}{384EI}$$

where

D = live load beam deflection in inches (same as Δy)
l = unbraced length of beam in inches
E = modulus of elasticity of the beam material in *psi*
I = moment of inertia of the beam shape in in^4

Using A36 steel we can substitute 29,000,000 psi for the value of E and $\delta(x)$, where $\delta = \frac{1}{360}$ and x is the center-to-center distance in inches between columns for the value of D. The design weight W for a given structure is either known or can be computed or estimated, and the unbraced length l of the beam is found using the procedure in Section 6.6. The only unknown item in the **standard deflection formula** is the value for I.

The **moment of inertia** I is a calculated value based on the shape of the crosssection of the beam. Larger moments of inertia indicate stronger and more stable shapes. (This is discussed in greater detail in Chapter 8.) If we know the required value for the moment of inertia of a beam, then we can find a suitable beam by looking in Table A.

Let us, then, rewrite the standard deflection formula so that we can use it to find a missing I value:

$$I_x = \frac{5Wl^3}{384ED}$$

Notice that the value for I now reads I_x. This is because the bending moment on the I-beam will occur against the x axis, since I-beams are usually oriented as shown in Figure 6.13. If you turn to any page in Table A where I-beams can be found, you will see a column for I_x and a column for I_y. Since the beams we will be studying will be assumed to be oriented with the web perpendicular to the ground we will use the I_x value when sizing our beams. You may also recognize that the value for I_x for any I-beam is significantly greater than its I_y value, which tells you that the beam loaded perpendicular to the x axis is far stronger.

With the above information in mind, let us look at an example of how the standard deflection formula is used in sizing an I-beam. Let us assume that an I-beam is needed to span two W14×34 columns located 12 ft O.C. oriented web-to-web, and that the beam must withstand a live load of 20,000 lb. To find the appropriate beam we use the following procedure.

1. Write the necessary formula:

$$I_x = \frac{5Wl^3}{384ED}$$

2. List all given information:

$$W = 20{,}000 \text{ lb}$$
$$E = 29{,}000{,}000 \text{ psi}$$

2a. Compute the allowable deflection:

$$\delta = \Delta y/x; \quad \delta = \tfrac{1}{360}, \; x = 144 \text{ in.}$$
$$D = \Delta y = \delta(x) = \tfrac{1}{360} \times 144 \text{ in.} = 0.4 \text{ in.}$$

3. Compute the actual length of the beam:

(*Note:* This step is done in this text for practice only. In actual conditions the column center-to-center length would be used, thereby insuring a built-in safety factor.)

Orientation: W-W (equal columns)
Formula: $B = l = L - 1 \text{ in.} - t_{w(W14\times34)}$
$L = 12 \text{ ft} = 144 \text{ in.}$
$t_{w(W14\times34)} = 0.285 \text{ in.}$
$B = l = 144 \text{ in.} - 1 \text{ in.} - 0.285 \text{ in.}$
$\underline{B = l = 142.715 \text{ in.}}$

Be sure to underline or circle your answer here to make it easier to find, since it will be used in the formula to compute I_x.

4. Substitute all known values into the I_x formula and compute the required value for I_x:

$$I_x = \frac{5(20{,}000 \text{ lb})(142.715 \text{ in.})^3}{384(29{,}000{,}000)(0.4 \text{ in.})}$$

(*Note:* From this point forward in this chapter we will not substitute the value for E since it is always the same.)

$$\therefore I_x = 65.26 \text{ in}^4$$

5. Select an appropriate beam from Table A. Since W14×34 type columns are being used we will try to find a beam that fits nicely against the flat part (*T*) of its web. The resulting assembly will look like the one in Figure 6.16. We, therefore, need a beam that has a b_f value less than or equal to the *T* value of the column and an $I_x > 65.26 \text{ in}^4$:

$$T_{(W14\times34)} = 12 \text{ in.}$$

Several options for the beam we might select based on the required I_x include

W8×21	$I_{x(ACT)} = 75.3$ in^4	$b_f = 5.270$ in.
W10×15	$I_{x(ACT)} = 68.9$ in^4	$b_f = 4.000$ in.
W12×14	$I_{x(ACT)} = 88.6$ in^4	$b_f = 3.970$ in.

Of these choices, the two on the bottom would be the best because they are smaller and lighter. The top beam would also fit against the W14×34 columns without cutting (discussed in Chapter 7) but weighs more per foot. Lighter beams tend to be less expensive since they require less material to manufacture, and they contribute less to the overall dead load of the structure.

Using the five-step procedure will help you tremendously in selecting an appropriate beam for a given live load. Please note that this procedure assumes that all beams are uniformly loaded with no significant load concentrated on any one portion of the beam. To size a beam for concentrated loads you would need to determine the shearing stress at the weakest point along the beam and size that beam based on its section modulus (S_x), a procedure that is not addressed in this text.

6.9 COMPUTATION OF MAXIMUM ALLOWABLE LIVE LOADS ON BEAMS

Using the standard deflection formula we can also determine the maximum live load that can be safely placed on an existing structure. By rewriting the deflection formula to solve for allowable load W we arrive at

$$W_{MAX} = \frac{384EI_xD}{5l^3}$$

This formula will yield the maximum uniform load allowed on a known beam in an existing structure.

To illustrate the use of this formula let us consider the following example. An S12×35.0 beam spans two S12×50.0 columns set 8 ft apart O.C. flange-to-flange. Compute the maximum allowable live load that can be placed on this beam.

1. Write the appropriate formula:

$$W_{MAX} = \frac{384EI_xD}{5l^3}$$

2. List all known information:

$$E = 29{,}000{,}000 \text{ psi}$$
$$I_{x(S12 \times 35.0)} = 229 \text{ in}^4 \quad \text{(found in Table A)}$$

2a. Compute the allowable live-load deflection:

$$D = \Delta y = \delta(x) = \tfrac{1}{360} \times 96 \text{ in.} = 0.27 \text{ in.}$$

3. Compute the actual length of the beam:

$$\text{Orientation: F-F (equal columns)}$$
$$\text{Formula: } B = l = L - 1 \text{ in.} - d_{(S12 \times 50.0)}$$
$$L = 8 \text{ ft} = 96 \text{ in.}$$
$$D_{(S12 \times 35.0)} = 12 \text{ in.}$$
$$B = l = 96 \text{ in.} - 1 \text{ in.} - 12 \text{ in.}$$
$$\underline{B = l = 83 \text{ in.}}$$

Be sure to underline or circle your answer here to make it easier to find since it will be used in the formula to compute W_{MAX}.

4. Compute the maximum allowable live load on the beam:

$$W_{MAX} = \frac{384 E (229)(0.27 \text{ in.})}{5(83 \text{ in.})^3}$$
$$= 240{,}837.54 \text{ lb}$$

This illustrates the incredible strength of the A36 steel I-beam!

6.10 SUMMARY

In this chapter we have looked at some of the properties of metals. We have seen that metals are malleable, ductile, and conductive. Steel is a strong human-made metal that is an alloy of iron and carbon. The use of steel in building structures makes them stronger and more durable. Without steel, it would be nearly impossible to build large structures and skyscrapers because few other materials could withstand the tremendous forces applied by such large structures.

We have also investigated some steel beam nomenclature and have learned about several important steel I-beam dimensions. Our study included the various ways that steel beams may be oriented within a building and touched upon the uses of steel cables and rods under tension. We have seen how uniformly distributed live loads affect the horizontal support members (beams) of a steel structure and have learned how to determine the load on a particular beam within a structure. Finally, we have discovered how beams may be sized and how their load-bearing capacity can be determined based on their allowed deflection value.

Now let us look at three practical application sample exercises that utilize the procedures covered in this chapter.

Sample Exercise 6.1

A steel loading platform is supported by 4 evenly spaced A36 steel rods 3 ft long each. The platform needs to have a loading capacity of 10,000 lb under less than 0.02 in. elongation. Compute the required diameter of the steel rods.

Using the basic formula for modulus of elasticity we have

Steel Beam Design

$$E = \frac{Pl}{Ae}$$

where

P = 10,000 lb/4 rods = 2500 lb/rod
l = 3 ft = 36 in.
e = 0.02 in.
E = 29,000,000 psi
A = ?

Rewriting this formula to solve for the missing area of cross section A, we find

$$A = \frac{Pl}{Ee}$$

Substituting our known values from above yields

$$A = \frac{2500 \text{ lb}(36 \text{ in.})}{E(0.02 \text{ in.})}$$
$$= 0.155 \text{ in}^2$$

This result, however, is not our final answer. To find the diameter of each of the rods we need to use the formula for the area of a circle $A = \pi r^2$. Rewriting this formula to solve for r, which is $\frac{1}{2}d$ (diameter), we have

$$r = \sqrt{\frac{A}{\pi}}$$

Solving for r we find

$$r = \sqrt{\frac{0.155}{\pi}}$$
$$= 0.394 \text{ in.}$$
$$\therefore d = 0.788 \text{ in.}$$

So each of the four rods would need to be 0.788 in. in diameter to meet the elongation and load criteria specified in this problem.

Sample Exercise 6.2

Determine an appropriate-sized beam for the beam labeled B in Figure 6.19.

First, we need to determine the load on beam B. As we know, it will carry one-fourth of the load from each of the two adjoining sections. Assuming uniform loads of 7000 lb/ft² for the left section and 14,000 lb/ft² for the right section, the sectional loads will be

Left section: 10 ft × 6 ft × 7000 lb/ft² = 420,000 lb
Right section: 4 ft × 6 ft × 14,000 lb/ft² = 336,000 lb
$W_B = \frac{1}{4}(420,000 \text{ lb}) + \frac{1}{4}(336,000 \text{ lb}) = 105,000 \text{ lb} + 84,000 \text{ lb} = \underline{189,000 \text{ lb}}$

FIGURE 6.19

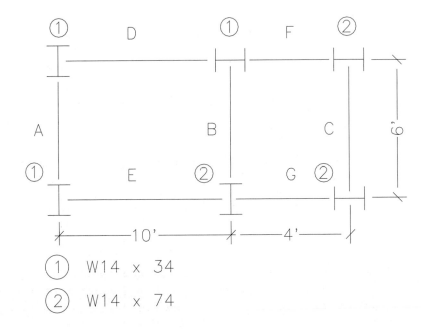

① W14 × 34
② W14 × 74

Now, we write the formula for I_x:

$$I_x = \frac{5Wl^3}{384ED}$$

Next, we write all known information:

$$W_B = 189{,}000 \text{ lb}$$
$$E = 29{,}000{,}000 \text{ psi}$$

Now, compute the allowable deflection:

$$D = \delta(x) = \tfrac{1}{360} \times 72 \text{ in.} = 0.2 \text{ in.}$$

Then, we compute the actual length of the beam:

Orientation: W-F (unequal columns)
Formula: $B = l = L - 1 \text{ in.} - t_{w(W14 \times 34)} - d_{(W14 \times 74)}$
$L = 6 \text{ ft} = 72 \text{ in.}$
$t_{w(W14 \times 34)} = 0.285 \text{ in.}$
$d_{(W14 \times 74)} = 14.17 \text{ in.}$ (Values for t_w and D found in Table A.)
$B = l = 72 \text{ in.} - 1 \text{ in.} - \tfrac{1}{2}(0.285 \text{ in.}) - \tfrac{1}{2}(14.17 \text{ in.})$
$\underline{B = l = 63.773 \text{ in.}}$

Now we substitute all of the known values into the I_x formula and compute the moment of inertia for this beam:

$$I_{xB} = \frac{5(189{,}000 \text{ lb})(63.773 \text{ in.})^3}{384E(0.2 \text{ in.})}$$

$$= 110.04 \text{ in}^4$$

Some of the smallest beam options are show below:

W8×31 $I_{x(ACT)} = 110$ in^4 $b_f = 7.995$ in.
W8×35 $I_{x(ACT)} = 127$ in^4 $b_f = 8.020$ in.
W10×22 $I_{x(ACT)} = 118$ in^4 $b_f = 5.750$ in.

Of these choices, the W10×22 would probably be the best selection since it is the smallest and lightest. The W8×31 has an $I_{x(ACT)}$ that is just a bit too low, and the W8×35 is the heaviest beam, making it a less likely choice.

Sample Exercise 6.3

For our final sample exercise let's consider the existing structure in Figure 6.20. We need to determine the maximum load capacity ($W_{MAX(Section)}$) for the entire 60-ft^2 section shown. To do this we must first find the load-bearing capacity of each of the lettered beams. Beams A and B are the same type beam spanning equal columns in the same orientation over the same span, so they will have the same load-bearing capacity. The same holds true for beams C and D. We will treat this problem as two separate beam problems and work them out individually. Let us begin with beams A and B.

1. Write the appropriate formula:

$$W_{MAX} = \frac{384EI_xD}{5l^3}$$

FIGURE 6.20

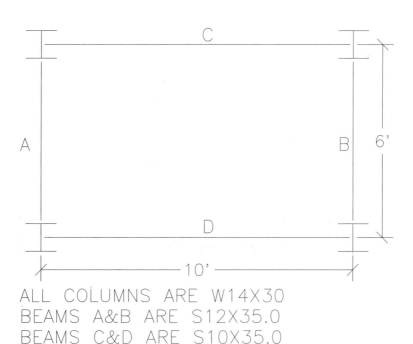

ALL COLUMNS ARE W14X30
BEAMS A&B ARE S12X35.0
BEAMS C&D ARE S10X35.0

2. List all known information:

$$E = 29{,}000{,}000 \text{ psi}$$
$$I_{x(S12\times 35.0)} = 229 \text{ in}^4 \quad \text{(found in Table A)}$$

2a. Now, compute the allowable deflection:

$$D = \delta(x) = \tfrac{1}{360} \times 72 \text{ in.} = 0.2 \text{ in.}$$

3. Compute the actual length of the beam:

Orientation: F-F (equal columns)
Formula: $B = l = L - 1 \text{ in.} - d_{(W14\times 30)}$
$L = 6 \text{ ft} = 72 \text{ in.}$
$d_{(W14\times 30)} = 13.84 \text{ in.}$
$B_{A\&B} = l_{A\&B} = 72 \text{ in.} - 1 \text{ in.} - 13.84 \text{ in.}$
$\underline{B_{A\&B} = l_{A\&B} = 57.16 \text{ in.}}$

4. Compute the maximum allowable live load on the beam:

$$W_{MAX(A,B)} = \frac{384E(229)(0.2 \text{ in.})}{5(57.16 \text{ in.})^3}$$
$$= 546{,}195.4 \text{ lb}$$

Now, let's find the maximum allowable load on beams C and D.

1. Write the appropriate formula:

$$W_{MAX} = \frac{384EI_xD}{5l^3}$$

2. List all known information:

$$E = 29{,}000{,}000 \text{ psi}$$
$$I_{x(S10\times 35.0)} = 147 \text{ in}^4 \quad \text{(found in Table A)}$$

2a. Now, compute the allowable deflection:

$$D = \delta(x) = \tfrac{1}{360} \times 120 \text{ in.} = 0.33 \text{ in.}$$

3. Compute the actual length of the beam:

Orientation: W-W (equal columns)
Formula: $B = l = L - 1 \text{ in.} - t_{w(W14\times 30)}$
$L = 10 \text{ ft} = 120 \text{ in.}$
$t_{w(W14\times 30)} = 0.270 \text{ in.}$
$B_{A\&B} = l_{A\&B} = 120 \text{ in.} - 1 \text{ in.} - 0.270 \text{ in.}$
$\underline{B_{A\&B} = l_{A\&B} = 118.730 \text{ in.}}$

4. Compute the maximum allowable live load on the beam:

$$W_{MAX(C,D)} = \frac{384E(147)(0.33 \text{ in.})}{5(118.73 \text{ in.})^3}$$

$$= 65{,}203.94 \text{ lb}$$

Now that we have the maximum allowable load on each of the beams, it would seem that all we have to do to find the maximum allowable load on the section is to add up all of the values. If we do this then we find

$$\Sigma W_{MAX(A-D)} = 2(546{,}195.4 \text{ lb}) + 2(65{,}203.94 \text{ lb})$$
$$= 1{,}222{,}798.7 \text{ lb}$$

Therefore, 1,222,798.7 lb is the maximum load that this section can carry, right? From Section 6.5 we know that each beam must carry one-fourth the total section load. Since $\frac{1}{4}(1{,}222{,}798.7 \text{ lb}) = 305{,}699.67$ lb, this will work great for beams A and B which have a capacity of 546,195.4 lb each. But beams C and D have a maximum load capacity of only 65,203.94 lb, which is far too weak to carry one-fourth of the total load shown above.

Just as a chain is only as strong as its weakest link, a section of a building is only as strong as its weakest element. We must, therefore, determine the maximum load-bearing capacity of this building section based on the strength of its weakest beam. The load-bearing capacity for this building section would be

$$\Sigma W_{MAX(A-D)} = 4(65{,}203.94 \text{ lb})$$
$$= 260{,}815.76 \text{ lb}$$

This figure is the **maximum section load capacity.** This structure, however, should not be used to carry a load this great. Instead, a safety factor should be determined. For example, if a 10% safety factor was used (discussed in Chapter 7), then this structure would have a design load-bearing capacity of a little over 230,000 lb. This value allows a reasonable safety margin for error in loading.

Problems

1. List three instances in which each of the following conditions might exist:

   (a) Oxidation
   (b) Thermal shock
   (c) Deformation
   (d) Flexure
   (e) Elongation
   (f) Permanent set

2. Name and explain some of the factors that might contribute to a structural live load.

3. Find any two metal objects and explain how, by their characteristics, you can tell that they are metal.

4. Place the items listed below in order from greatest to least value for modulus of elasticity:

   (a) Wood pencil
   (b) Marshmallow
   (c) Iron rod
   (d) Rubber band
   (e) Chalk stick
   (f) Plastic spoon

5. Compute the deflection ratio value δ for each of the following beams:

Horizontal Span	Vertical Bend
(a) 6 ft	0.5 ft
(b) 8 ft	0.3 ft
(c) 12 ft	2 in.
(d) 15 ft	$2\frac{1}{2}$ in.
(e) 20 ft	$\frac{3}{4}$ in.
(f) 9 ft 3 in.	$\frac{1}{4}$ in.
(g) 21 ft $4\frac{3}{8}$ in.	$\frac{1}{32}$ in.

6. Do any of the ratios in Problem 5 meet ASTM standards for steel beam live-load deflection?

7. How many $\frac{1}{8}$-in.-diameter steel threads 3 ft long each would need to be braided together to support a load of 10,000 lb if 0.06 in. of elongation is permitted?

8. Compute the modulus of elasticity for each of the following 4-ft-long 1-in.-diameter supports carrying a load of 4 kips if the measured elongation of the support is

   (a) 0.01 in.   (d) 0.5 in.
   (b) 0.03 in.   (e) 1 in.
   (c) 0.1 in.

9. What might be the materials used in each of the supports in Problem 8?

10. A loading hook is mounted by 24 steel bolts to a main assembly frame. If the bolts are 3/8 in. diameter and 1 in. long each and the longitudinal tolerance when stressed may not exceed 0.015 in., compute the maximum load capacity of the hook assembly.

11. Compute the load on each lettered beam in Figure 6.21.

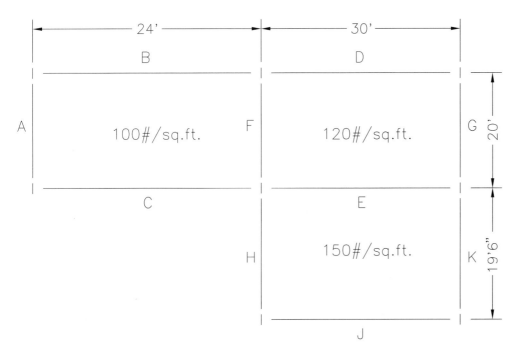

FIGURE 6.21

12. Recalculate the loads on beams F and H in Problem 11 if additional columns are located 10 ft away from the column that joins these two beams along each beam.

13. Using Table A, fill in the missing information for each beam listed in the chart below:

Beam name	Shape	Design depth	Wt/lf	d	t_w	I_x
(a) W14×74						
(b) W18×106						
(c) W12×50						
(d) S12×35						
(e) WT6×29						
(f) MC8×20						
(g) ST6×25						
(h) L8×8×$\frac{5}{8}$						

14. Compute the actual cut length of each beam spanning the columns listed in the chart below:

	Column 1	Column 2	Orientation	Distance (L)
(a)	W14×34	W14×34	W-W	10 ft
(b)	W14×30	W14×30	W-F	12 ft
(c)	W14×74	W14×74	F-F	16 ft
(d)	W12×50	W14×22	W-F	9 ft
(e)	W12×50	W14×22	F-W	9 ft
(f)	W12×50	S12×35	F-F	9 ft 9 in.
(g)	S12×31.8	S10×35	W-F	7 ft $9\frac{1}{4}$ in.
(h)	S10×35	S12×31.8	W-F	7 ft $9\frac{1}{4}$ in.
(I)	HP12×74	W14×74	W-W	16 ft $2\frac{1}{2}$ in.
(j)	HP10×57	HP10×57	F-F	17 ft $6\frac{1}{8}$ in.

15. Refer to Figure 6.21. Compute the cut length of each lettered beam using W14×34 columns.

16. Compute the moment on inertia for each beam in Problem 15. Using your results, select an appropriate beam from each of the following categories for each lettered beam using Table A:

    (a) W12 (d) S12
    (b) W10 (e) S10
    (c) W8

17. Refer to Figure 6.21. Ignore the loads shown in the diagram and do the following:

    (a) Assume that all beams in the diagram are W14×30 and all columns are W12×50. Compute the cut length of each beam.
    (b) Using Table A, find the moment of inertia for each beam.
    (c) Compute the maximum allowable load on each beam using the information from parts (a) and (b) of this problem.

18. Compute the maximum allowable load for each section of the building in Problem 17.

Steel Joist and Fastener Design

7.1 INTRODUCTION

This chapter introduces you to the basics of steel joist and fastener design and detailing. The following topics are covered:

1. Steel framing plans
2. Sizing of
    (a) Steel beams as joists
    (b) Bolted/riveted beam clips
    (c) Bolts and rivets
3. Computation of load safety factors
4. Detailing of
    (a) Steel beams
    (b) Steel beam clips
5. Custom steel beam cutting

This picture shows a steel-framed building under construction. From this view all of the structural members are exposed. The columns and main beams are clearly visible and the joists shown are resting on top of the beams. Notice the small, staggered pieces of metal between the joists. They are called spreaders and their purpose is to maintain an even distance between the joists. All of these elements are indicated on a steel framing plan. *Courtesy of MML Construction.*

FIGURE 7.1

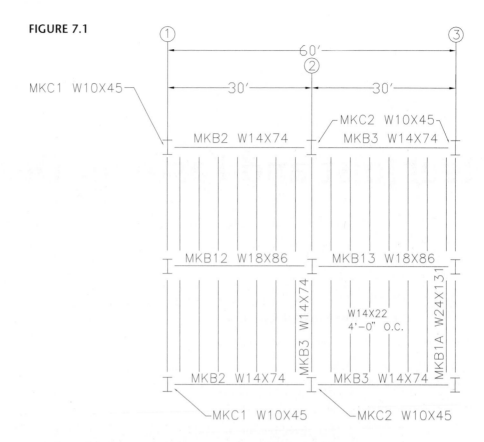

7.2 STEEL FRAMING PLANS

Figure 7.1 is an illustration of a **steel framing plan**. The single lines labeled with the letters MK, meaning "mark," represent the beams and spreaders (which are discussed in the next section). The I shapes represent the I-beam columns of the structure. Notice that every element of the plan has been labeled. This is necessary because every beam and column must be prepared to meet specific mounting and assembly requirements prior to being sent on site for construction. Throughout this chapter we will be looking at the detailing of the horizontal structural members and their fasteners only. It is, therefore, important to familiarize yourself with the information on steel framing discussed in this section.

All necessary dimensions for constructing a steel superstructure are shown on a steel framing plan. The dimensions shown on the plan in Figure 7.1 fall into four categories:

1. *Overall dimensions:* Show the distance between the outermost features of the plan.
2. *Column line dimensions:* Show the center-to-center distance between columns.
3. *Spreader/joist spacing:* Is called out as a note over a double-sided arrow and indicates the standard spacing on-center between spreaders or joists.

4. *Miscellaneous dimensions:* Used to show detailed dimensioning of any features not called out by the three dimensioning methods shown above.

Column lines are used to indicate locations of various features on the plan. These lines indicate the center lines of the columns on the plan. By standard convention, numbers and letters are used to designate the individual column lines and they will appear on opposite axes. For example, since the vertical column lines in Figure 7.1 are labeled using numbers, the horizontal column lines are labeled using letters.

Beam marks are used to tell the manufacturer how to mark the beam before it is sent to the construction site. The letters MK will appear, followed by an alphanumeric designation, such as B1A. The note MKB1A appearing above a beam on a steel framing plan indicates to the manufacturer that this beam must be spray painted or stenciled with the letters B1A.

In drawing the framing plan all MK labels go on top of horizontal beams and to the left of vertical beams. The beam nomenclature excluding the beam length is added following a space after the MK designator. These labels are discussed further in Section 7.9 along with the preparation of steel beam cutting lists.

Note: The designation "C" is usually reserved for columns. Therefore, a feature labeled "MKC1B W14×74" would be a column designation.

7.3 SIZING JOISTS

A **joist** is a steel beam that spans two other horizontal members or beams. It is a **secondary structural member** because it does not directly transfer the load it carries to the foundation of the structure. This is in contrast to a **beam**, which is a **primary structural member** because it is directly attached to the foundation by two or more columns.

These joists have been fastened to the beams shown and braced using angled braces. This is just one of many ways in which joists may be attached to a steel structure. *Courtesy of MML Construction.*

The method for sizing joists is very similar to the method for sizing individual steel beams, with the exception that many joists will carry a given load. Deflection and moment of inertia are still computed in the same fashion as for single beams, but the load must be divided evenly among the joist beams. Again, we will assume, as in Chapter 6 that all loads are uniformly distributed among the members. To examine how this is done let us consider the following problem.

Using the diagram from Sample Exercise 6.3 and a live load of 1500 lb, we will run 4 joists at 2 ft O.C., as shown in Figure 7.2. To size these joists we need to know the same information as we would to size single beams, so we can use the procedure to size steel I-beams by moment of inertia.

1. Write the necessary formula:

$$I_x = \frac{5Wl^3}{384ED}$$

2. List all given information:

$$W/\text{joist} = 1500 \text{ lb}/4 \text{ joists} = 375 \text{ lb/joist}$$
$$E = 29,000,000 \text{ psi}$$

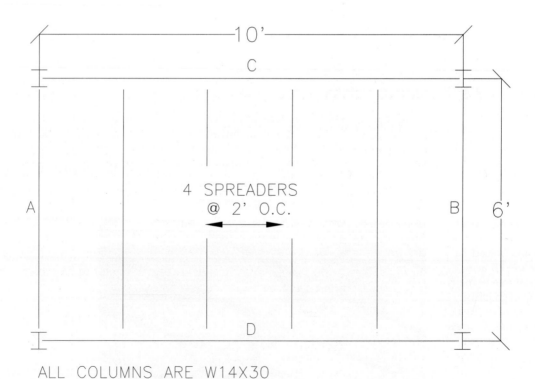

FIGURE 7.2

2a. Compute the allowable deflection:

$$\delta = \Delta y / x \quad \delta = \tfrac{1}{360}, \; x = 72 \text{ in.}$$
$$D = \Delta y = \delta(x) = \tfrac{1}{360} \times 72 \text{ in.} = 0.2 \text{ in.}$$

3. Compute the actual length of the beam. This step may be omitted and the center-to-center distance may be used. This will build in a small safety factor for the joists.

4. Substitute all known values into the I_x formula and compute the required value for I_x:

$$I_x = \frac{5(375 \text{ lb})(72 \text{ in.})^3}{384(29{,}000{,}000)(0.2 \text{ in.})}$$
$$= 0.314 \text{ in.}^4$$

5. Select an appropriate beam from Table A at the back of this text. Our selection for this joist is a WT2×6.5, which has an $I_{x(\text{ACT})} = 0.526$ in.^{4} This beam will need to be custom cut to fit against the web of the S10×35's into which it is being run. This procedure is outlined and explained in Section 7.8.

7.4 CLIP SELECTION

Once all of the columns, beams, and joists have been selected for a given structure, a method must be determined to hold it all together. In this text we discuss the use of **beam clips** and bolts/rivets to fasten these structural elements together. We must be certain that all of the components in a structure will have the proper structural integrity. This includes the devices used to fasten these items together. Table B in the back of this text lists 16 different sizes of beam clips that are designed to be used with bolts or rivets.

Figure 7.3 shows two W14×34 beams attached to a W16×50 column. Notice the configuration of the clip shown. This is a 19-lb clip. Its dimensions are 4 in. × $3\tfrac{1}{2}$ in. × $\tfrac{3}{8}$ in. × 0 ft $11\tfrac{1}{2}$ in. (found on page 293 in Table B). A clip is always identified by its weight, including the weights of its rivets or bolts. The dimension nomenclature for clips is shown below:

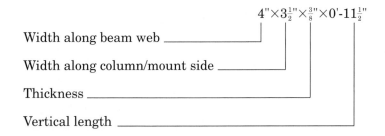

FIGURE 7.3

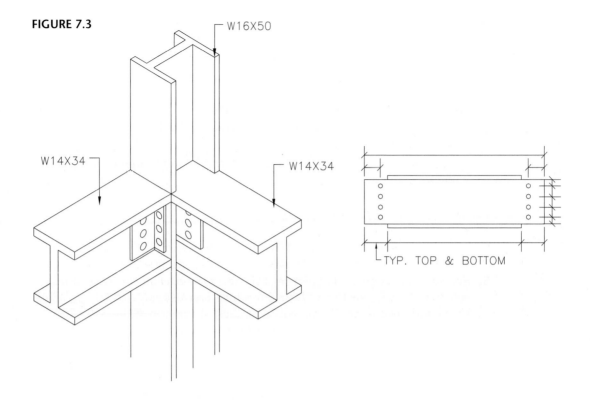

The method for selecting beam clips is really quite simple. There are only three criteria:

1. Beam/joist size
2. Column size (to determine maximum clip size)
3. Single/double shear of clip assembly (see Section 7.5)

In Figure 7.3 the 19-lb clip was selected for two reasons:

1. The size of the beam is W14×34 and a 19-lb clip is required for a nominal 14-in. beam (14 in B).
2. The size of the column is W16×50 and a maximum 19-lb clip is allowed for a nominal 16 in. column under 50 lb/LF (16 in C 50 lb and under).

In the next section we investigate the shear strength of the bolts/rivets so that we can be sure to use bolts or rivets large enough to carry the expected load. Remember that the clip size identified by the column used denotes the *maximum* clip size that is allowed on that column. Smaller clips may be used as needed.

Notice that all the clips shown in Table A have an array of rivet/bolt holes. These holes must be punched in both the clip and the beam at the designated locations. We do not discuss column detailing in this text, but the concept is the same for both beams and columns. In Section 7.6 we see how these bolt/rivet holes are referenced and dimensioned. Now, we investigate the shear strength of the clips required for a given assembly.

7.5 BOLT/RIVET SHEAR

No matter how strong the structure is that we design, it is no good if the fasteners we use to hold it together will not support the loads adequately. We, therefore, need to determine the size of bolt or rivet required to mount a particular clip. The method we use for doing this is as follows

1. **Single shear** occurs when two surfaces that have forces acting in opposite directions are riveted together. We determine the amount of single shear on the mounting side of the clip. For example, the shear strength of a steel rivet is 10,000 lb/in^2. To determine the single shear of a riveted or bolted surface we must multiply the shear strength of steel by the cross-sectional area of the rivet/bolt and also multiply that value by the number of rivets/bolts used. This yields

$$S = 10{,}000 \text{ lb/in}^2 \times AN$$

where

A = cross-sectional area of the rivet in in^2
N = number of rivets used

Substituting $\frac{1}{4}\pi d^2$ for A we find

$$S = 10{,}000 \text{ lb/in}^2 \times \tfrac{1}{4}\pi d^2 N$$
$$d = \text{diameter of a rivet in inches}$$

By isolating d we can calculate the minimum rivet size for a given load:

$$d = \sqrt{\frac{S}{2500\pi N}}$$

where

S = expected load on beam joint
N = number of rivets at joint

This formula will be used to compute the size of rivets needed on the mounting side of the clip.

Note: The value for shear S will be one-half the total load on a beam mounted at either end. This is the value that will be used throughout this text.

---

2. To determine the size of rivet/bolt needed on the web side of the clip we need to use the formula for **double shear** of the fastener. We generate this formula by substituting twice the shear value in the original formula to arrive at

$$d = \sqrt{\frac{S}{5000\pi N}}$$

where

S = expected load on beam joint
N = number of rivets at joint

Using the two formulas derived above we can determine the size of rivet/bolt needed to mount a chosen clip at its web and against its support. For our first example we will compute the size of rivets needed to mount one of the W14×34 beams shown in Figure 7.3 if that beam is carrying a 10,000-lb load.

First, we will find the size of rivets needed to attach the clip at the column. For this we will use the single shear formula:

$$d = \sqrt{\frac{S}{2500\pi N}}$$

where

$S = \frac{1}{2}(10{,}000 \text{ lb}) = 5000 \text{ lb}$
$N = 4$ rivets on a 19-lb clip

$$d = \sqrt{\frac{5000 \text{ lb}}{2500\pi(4 \text{ rivets})}}$$

$$= 0.399 \text{ in.}$$

Therefore, we would use $\frac{1}{2}$-in. rivets.

Note: Standard bolts and rivets come in increments of $\frac{1}{8}$ in. Therefore, it will usually be necessary to size up the value found in the shear formula to the next highest bolt or rivet size.

Next, let's find the size of the rivets needed to fasten the web side of the clip to the beam web. We need to use the double shear formula:

$$d = \sqrt{\frac{5000 \text{ lb}}{5000\pi(4 \text{ rivets})}}$$

$$= 0.282 \text{ in.}$$

Therefore, we would use $\frac{3}{8}$-in. rivets.

Now that we have seen how all of the features of the fasteners are found we are ready to move on to the method used to detail this information on a drawing.

This walking bridge is supported using an interesting structural design concept. The arched portion in the center is actually a support for the main section of the bridge. As you can see from the close-up views, horizontal members have been built into the arch upon which the main part of the bridge can rest.

7.6 COMPUTATION OF LOAD SAFETY FACTORS

In practical field application a **safety factor** (s.f.) is usually included in all structural designs. The principle of design load safety factors falls into two categories: oversizing and overloading.

1. **Oversizing:** This procedure is used to design a structural element to be stronger than it is actually required to be. We will use this application when designing a new structure. The oversizing safety factor may be applied to the computed design load or it may be applied to the beam moment of inertia I. The simple oversizing formula is

$$W_{s.f.} = W(1 + s.f.)$$

where

$W_{s.f.}$ = load with safety factor included
W = original design load
s.f. = safety factor in decimal form

or, for oversizing based on moment of inertia,

$$I_{s.f.} = I(1 + s.f.)$$

where

$I_{s.f.}$ = moment of inertia with safety factor included
W = original moment of inertia
s.f. = safety factor in decimal form

2. **Underloading:** This procedure is applied to structures that already exist and require no additional structural changes. In this procedure the maximum load capacity W_{MAX} of an existing structure is computed, then that value is reduced by the underloading safety factor. The simple underloading formula is

$$W_{s.f.} = W(1 - s.f.)$$

where

$W_{s.f.}$ = load with safety factor included
W = original design load
s.f. = safety factor in decimal form

To illustrate these principles, let us look at two examples involving the application of safety factors.

Example 1

After evaluating the specifications for a particular steel beam, you compute its required moment of inertia (I_x) to be 374 in⁴. You then select a W12×50 beam for this load ($I_{x(ACT)}$ = 394 in⁴). Your client then decides that this structure should have a built-in safety factor of 40%. Resize the steel beam based on this criteria. Since the moment of inertia is already known we can simply use the oversizing formula based on the computed value of I_x:

$$I_{s.f.} = I(1 + s.f.)$$
$$I_{s.f.} = 374 \text{ in}^4(1 + 0.40)$$
$$I_{s.f.} = 523.6 \text{ in}^4$$

Therefore, a W14×53 would be an appropriate selection ($I_{x(ACT)}$ = 541 in⁴).

Example 2

An existing structure was originally designed to carry a maximum load of 48,500 lb. Due to age and use it has been determined that the structure is somewhat less sound than it had been immediately following its construction. You must reevaluate the structural integrity of this structure using an underloading safety factor of 35%. The formula for computing the underloading value of an existing structure is

$$W_{s.f.} = W(1 - s.f.)$$

Substituting the current values for W and s.f. we find

$$W_{s.f.} = 48,500 \text{ lb}(1 - 0.35)$$
$$= 31,525 \text{ lb}$$

This represents the maximum allowable load including the 35% s.f.

STEEL JOIST AND FASTENER DESIGN   161

7.7 CLIP DETAILING

Once you have selected a clip or clips to be used in the assembly of a structure, all specifications pertaining to that clip or clips must be recorded in both chart and drawing form. Locations and sizes of bolt or rivet holes are essential pieces of information, since holes punched in the beams and columns must match exactly the locations of holes punched in the clips.

Figure 7.4 shows the complete isometric detailing of a 19-lb clip. Most of the information shown can be found in Table B in the back of this text. The value for g, however, which is the distance to the centerline of the holes on the mount side of the clip measured from the back of the clip, can be determined only once the web thickness of the beam to be hung is known.

The distance from the back of the clip to the centerline of the clip holes is called the **gage distance**. The centerline that runs in any direction through a hole or row of holes parallel to any one of the planes of the clip is called a **gage line** (see Figures 7.4 and 7.5). The distance between gage lines measured along the length of the clip is called the clip **pitch**. This distance indicates the standard O.C. spacing between holes along the clip.

Since a clip is really a segment of an angled beam, it is designated on most details with the angle symbol, ∠. A clip is individually detailed on a steel beam detail as shown in Figure 7.6. The following nomenclature is used when detailing a steel beam clip.

FIGURE 7.4

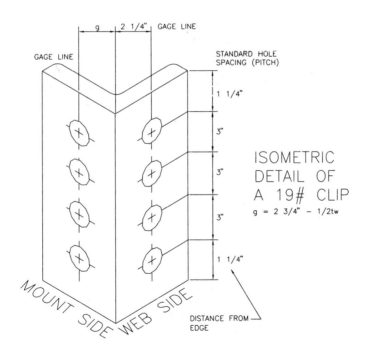

FIGURE 7.5

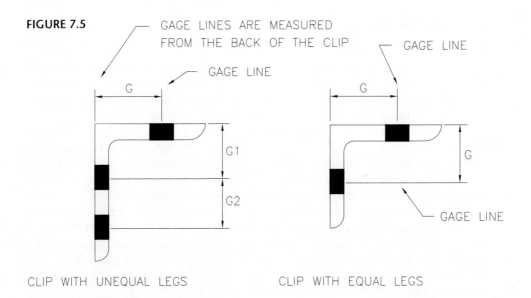

CLIP WITH UNEQUAL LEGS CLIP WITH EQUAL LEGS

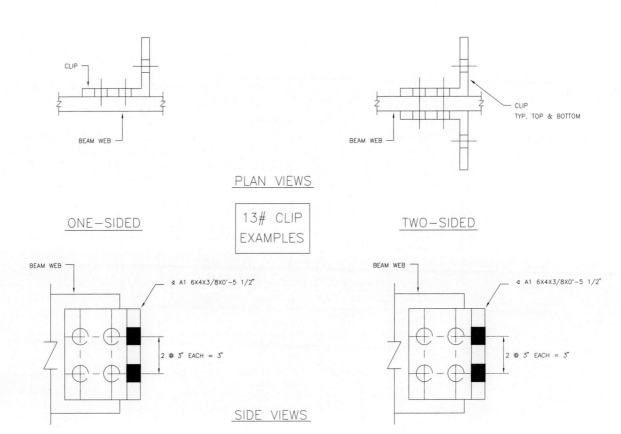

FIGURE 7.6

Clip dimensions—Contain the clip nomenclature, including information on how many sides of the beam the clips are located. The breakdown of the clip shown in Figure 7.6 as follows:

$$2\angle s\ A1\ 6\times4\times\tfrac{3}{8}\times0'-5\tfrac{1}{2}''$$

- Number of clips at this location
- Arbitrary clip chart designation
- Width along beam web
- Width along column/mount side
- Thickness
- Vertical length

Gage hole dimensions—Show the number of rivet/bolt holes, their spacing, and total O.C. length along the mount side of the clip. An explanation of the gage hole nomenclature for the 13-lb clip in Figure 7.6 is given below:

$$1@3''=3''$$

- Number of spaces between holes
- Pitch distance
- Total O.C. distance

Clip list—Contains all of the information about each clip used in a structure. These lists are generally used only on larger-scale projects. When a clip list is used the clip nomenclature may be omitted from the clip dimension information since it will appear in the list. Clip lists are similar to bills of materials or schedules. Some may also contain a column for quantity of clips to be used, while others may not. An example of a clip list is shown in Table 7.1.

It is acceptable for you to use any combination of the items listed in this section to adequately and accurately detail the clips in any project you work on. Always be sure that all of the required information, especially that which pertains to dimensioning, is included in your details and charts.

In Section 7.9 we will see how the information discussed in this section is inserted into a complete steel beam detail drawing. First, however, there is one more item that we need to address.

TABLE 7.1

| | | Angle Clips | | | |
| | | | Clip Length | | |
Designation	Size (lb)	Dimensions	Feet	Inches	Rivet size
a1	13	$6\times4\times\tfrac{3}{8}$	0	$5\tfrac{1}{2}$	$\tfrac{1}{2}$
a2	19	$4\times3\tfrac{1}{2}\times\tfrac{3}{8}$	0	$11\tfrac{1}{2}$	$\tfrac{3}{8}$
a3	25	$4\times3\tfrac{1}{2}\times\tfrac{3}{8}$	1	$2\tfrac{1}{2}$	$\tfrac{3}{8}$

7.8 CUSTOM BEAM CUTS

In Figure 7.7 we see four ways that the end of an I-beam may be cut in order to fit against the web of another beam or column. Two of these methods, the cope and the block, are used at the ends of joist beams to allow the web of the joist to meet the web of the main beam. Figure 7.8 clearly shows a joist attached to a main beam. The other two methods used for cutting steel I-beams, the cut and cut and chip, are used where main beam members join to the web of an I-beam column. Each of these methods is discussed below.

Cope

The joist in Figure 7.8 has been cut using a cope. As shown in Figure 7.7 in a cope the flange of the joist is removed down to the web at the point where the fillet meets the flat portion of the web. The depth to which the cope is cut is determined by the size of the joist, not the size of the main beam.

The length of the cope, or the distance that the joist flange is cut back away from the main beam flange, is equal to one-half the a value of the main beam plus any required or called-out tolerance(s). Since a value of $\frac{1}{2}$-in. tolerance is already built in to the beam length formulas (see Section 6.7), no additional tolerance will be used for the remainder of this section.

The dimensions needed for correctly detailing a cope are shown in Figure 7.9. The top view of the cope is shown in this diagram but is not generally needed because a cope has no inset dimension along the flange.

FIGURE 7.7

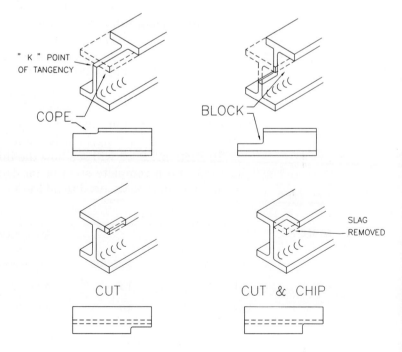

Steel Joist and Fastener Design   165

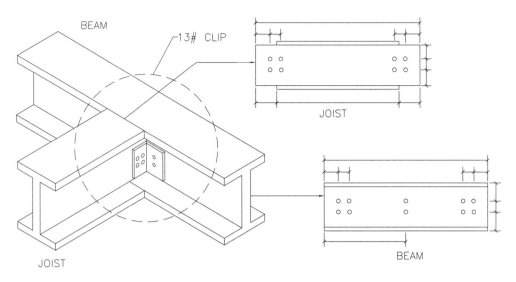

FIGURE 7.8

FIGURE 7.9

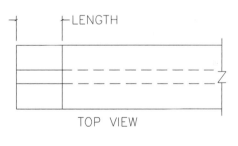

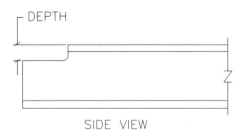

COPE DETAIL

Block

The **block** is the other type of cut that may be used where a joist joins to a main beam. The depth of a block may be up to one-half of the total joist depth. A block is used wherever a large portion of the joist must be removed in order for it to be joined to a main beam. The length of a block is determined in exactly the same fashion as the length of a cope.

The dimensions needed for correctly detailing a block are shown in Figure 7.10. The top view of the block is shown in this diagram but is not generally needed because a block has no inset dimension along the flange.

Cut

Figure 7.11 shows an illustration of a **cut** in a steel beam. Notice that a portion of the flange has been trimmed away from the end of the beam that meets the web of the column. A cut only removes a portion of the beam flange and does not affect the web at all. If the beam is to be cut at both sides, as shown in Figure 7.11, then the inset of the cut (distance measured along the flange breadth of the beam) is computed by subtracting one-half the value of T for the column from one-half the flange breadth b_f of the beam. If the beam flange is to be cut along one side only, then the inset will be found by subtracting the T value of the column from the b_f value of the beam.

Note: The inset of a beam cut may not exceed the a value of the beam minus the fillet radius R. This will place a restriction on the size of a single-sided cut allowed on any particular beam.

FIGURE 7.10

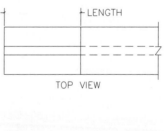

TOP VIEW

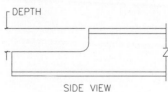

SIDE VIEW

BLOCK DETAIL

FIGURE 7.11

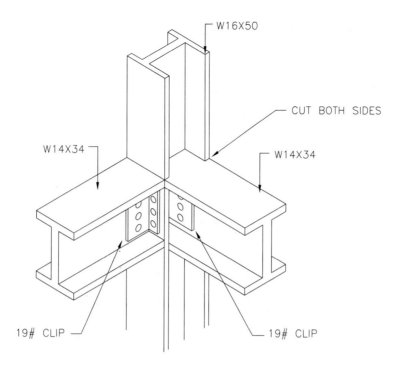

The length of the cut is equal to one-half the a value of the column.

The dimensions needed for correctly detailing a cut are shown in Figure 7.12. The side view of a cut is shown in this diagram but is generally excluded because a cut involves no modification to the web of the beam.

Cut and Chip

A **cut and chip** is practically identical to a cut except that the inset is equal to the a value of the beam and is not dependent on any of the column dimensions. A cut and chip literally strips away all of the flange along a portion of a beam (see Figure 7.7).

The length of a cut and chip is found just like the length of a cut. The dimensions needed for correctly detailing a cut and chip are shown in Figure 7.13. The side view of a cut and chip is shown in this diagram but is generally excluded because a cut and chip involves no modification to the web of the beam.

FIGURE 7.12

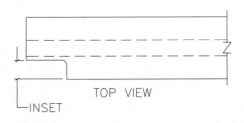

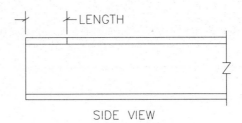

FIGURE 7.13

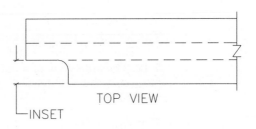

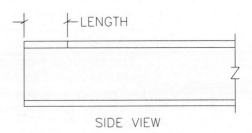

7.9 ASSEMBLING THE STEEL BEAM DETAIL

Once all of the items discussed in this chapter have been determined they need to be assembled and inserted into the final drawing product called the **steel beam detail** drawing. In this section we will discuss the assembly of this type of drawing. The following items appear on a steel beam detail.

Beam mark—A designation such as B1. Appears on any dimension that typically applies to that beam.

Beam length—Indicates the cut length of a beam as discussed in Chapter 6.

Number req'd—Lists the number of a particular type of beam needed for a structure.

Holes—Bolt/rivet holes are shown exaggerated in size but centered on their designated location. Hole sizes may be called out or indicated in a chart.

Clips—Clip dimensions, locations, and lists are used to indicate the specifications for each clip used in a structure.

Beam cuts—Any custom cuts required are shown and detailed on the steel beam detail. Cut locations are also noted by the following indicators:

T—Top flange
B—Bottom flange
NS—Near side of beam
FS—Far side of beam
BE—Both ends of the beam

Cutting list—A complete cutting list containing all of the beams detailed on a particular drawing is included.

Figure 7.14 illustrates a typical steel beam detail drawing on which four similar steel beams are shown. Four length dimensions are shown, one each for beams F10 through F13. Each horizontal dimension includes the beam name (F10, etc.), the number required for the structure (e.g., "4 REQ'D"), and the complete beam nomenclature, including the actual cut length in feet and inches.

Horizontal dimensions on a steel beam detail drawing are not typically to scale, but are shortened to accommodate the size of the drawing. Above the beam you will notice the dimensions for the locations of the bolt/rivet holes. Hole sizes may be indicated by leader notes or general notes, and do not appear to scale on the drawing. Because of their relatively small size, holes are shown larger than they would actually appear in the drawing scale.

Clips are shown at the ends of the beam. The clips are detailed according to the criteria discussed in Section 7.7. A clip list may be included if the drawing details two or more clips. In Figure 7.14 no clip list is included because only one clip, a 30-lb clip, is shown.

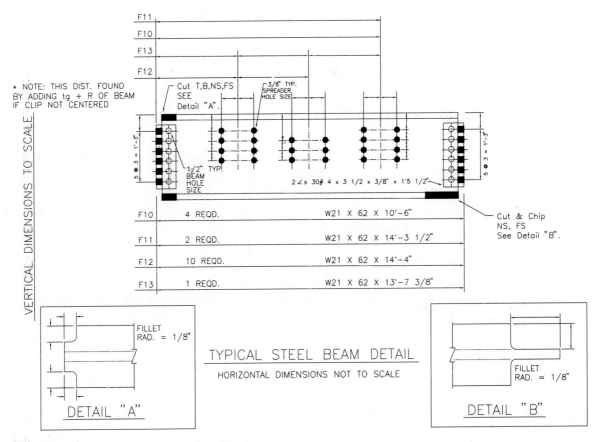

FIGURE 7.14

The custom beam cuts must also be detailed on the steel beam detail drawing. The beam shown in Figure 7.14 has two different types of custom cuts. Four cuts are needed on the left side of the beam. The leader note above the top cut explains the locations of all of the required cuts (top, bottom, near side, far side) and refers to detail A, which shows the dimensions of the cut. The lower right end of the beam has two cut and chips as shown (near side, far side) and refers to detail B.

More than one complete beam detail may appear on a steel beam detail drawing if necessary. It is, however, good practice to prevent overcrowding of details since these drawings can become quite complicated.

7.10 SUMMARY

In this chapter we have seen all of the basic elements required to effectively assemble a steel beam detail drawing. This drawing is designed to accompany a steel framing plan on which the centerlines of all of the steel members of a structure are shown. We also looked at the sizing and mounting of joist beams, which span two horizontal structural members called main load-bearing beams. The selection of fasteners is also critical to the structural integrity of a system. We have discussed how to accurately select clips and size the bolts/rivets used to hold these in place. Now, let us look at a couple of sample exercises that illustrate some of the practical applications of the material discussed in this chapter.

Sample Exercise 7.1

A 20-ft × 20-ft building section carries a load of 60 psf (lb/ft^2). The structure requires joists to be placed 4 ft O.C. Determine the size of joists needed to support the load in this section.

We begin this problem by determining the number of joists to be used. Since the joist beams will begin at 4 ft from the first main beam and continue every 4 ft to a distance of 16 ft we will need four joists here (see Figure 7.15). Next, we need to determine the load on each joist, assuming uniform load distribution. The load on the entire section is

$$20 \text{ ft} \times 20 \text{ ft} \times 60 \text{ psf} = 24{,}000 \text{ lb}$$

This will make the load per joist

$$24{,}000 \text{ lb} \div 4 \text{ joists} = 6000 \text{ lb}$$

FIGURE 7.15

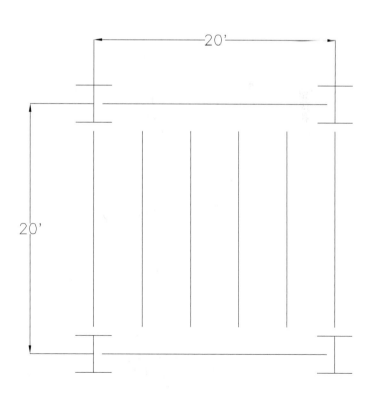

Now, we need to find the allowable deflection of a joist so that we can size it using the standard deflection formula:

$$D = \tfrac{1}{360} \times 240 \text{ in.}$$
$$= 0.667 \text{ in.}$$

Next, we insert the above value into the standard deflection formula to find the required moment of inertia for a joist:

$$I_x = \frac{5(6000 \text{ lb})(240 \text{ in.})^3}{384E(0.667 \text{ in.})}$$

$$= 55.58 \text{ in}^4$$

Some options for this joist include

S8×23 $I_{x(ACT)} = 64.9$ in.^{4}
W10×15 $I_{x(ACT)} = 68.9$ in.^{4}
W8×18 $I_{x(ACT)} = 61.9$ in.^{4}

The final beam selection will be based on criteria discussed in Chapter 6.

Sample Exercise 7.2

A steel beam carries a uniform load of 45,000 lb and is fastened using standard clip orientation at either end with 24-lb clips. Compute the standard size rivets needed for this job.

The load at one end of the beam would be

$$S = 45{,}000 \text{ lb} \div 2 = 22{,}500 \text{ lb}$$

The number of rivets needed (see Table B) is

$$N_{WEB} = 6$$
$$N_{MOUNT} = 12$$

The web side rivet size is computed using double shear:

$$d = \sqrt{\frac{S}{5000\pi N}}$$

$$= \sqrt{\frac{22{,}500 \text{ lb}}{5000\pi(6)}}$$

$$= 0.489 \text{ in. or } \tfrac{1}{2}\text{- in. rivets}$$

The mount side rivet size is computed using single shear:

$$d = \sqrt{\frac{S}{2500\pi N}}$$
$$= \sqrt{\frac{22{,}500\,\text{lb}}{2500\pi(12)}}$$
$$= 0.489 \text{ in. or } \tfrac{1}{2}\text{- in. rivets}$$

Both sides will require $\tfrac{1}{2}$-in. rivets.

Sample Exercise 7.3

Assume that the beams in Sample Exercise 7.1 are W10×19's and the columns are W10×49's, using the W10×15's found in the same exercise as joists. Detail each horizontal member of the structure.

From Table B in the back of this text we find that a 13-lb clip is to be used on 10-in. beams. Therefore, the clip detailing information will read

$$13\# \ 6 \times 4 \times \tfrac{3}{8} \times 0' - 5\tfrac{1}{2}''$$

Next, we will compute the length of the load-bearing beams and joists. The web-to-web spanning beam length will be

$$B_{\text{W-W}} = L - 1 \text{ in.} - t_{w(W10\times49)}$$
$$= 240 \text{ in.} - 1 \text{ in.} - \tfrac{5}{16} \text{ in.}$$
$$= 19 \text{ ft } 10\tfrac{11}{16} \text{ in.}$$

The flange-to-flange spanning beam length will be

$$B_{\text{F-F}} = L - 1 \text{ in.} - d_{(W10\times49)}$$
$$= 240 \text{ in.} - 1 \text{ in.} - 10 \text{ in.}$$
$$= 19 \text{ ft } 1 \text{ in.}$$

The joist length will be calculated using the web-to-web formula for spanning between two equal columns using the main beam value for web thickness:

$$B_{\text{Joist}} = L - 1 \text{ in.} - t_{w(W10\times19)}$$
$$= 240 \text{ in.} - 1 \text{ in.} - \tfrac{1}{4} \text{ in.}$$
$$= 19 \text{ ft } 10\tfrac{3}{4} \text{ in.}$$

Now we need to determine the types and dimensions of any custom cuts to the ends of the beams and joists. Since the b_f value of the W10×19 main beams is 4 in., which is less than both the T and d values of the W10×49 columns (see Figure 7.16), these beams will require no cuts. Additional detailing of these beams is discussed shortly. First, however, we also need to determine if the joists will require any custom cuts. Figure 7.17 shows the end view of one of the joists as it meets a beam. Notice that the top of the joist is level

FIGURE 7.16

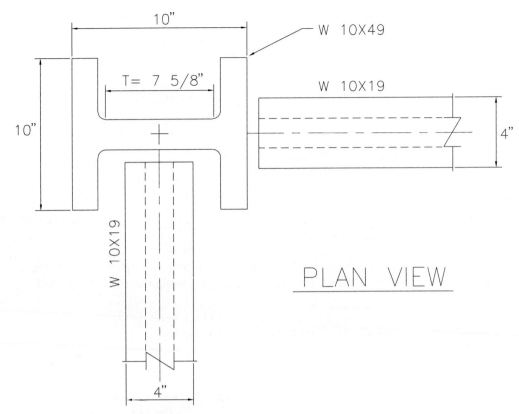

PLAN VIEW

FIGURE 7.17

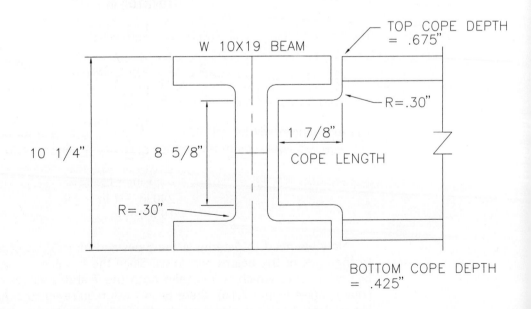

END VIEW

with the top of the beam. This will be the starting point from which all of the cutting will be measured.

The top of the joist must be cut down through the flange to a depth equal to $t_f + R$ of the W10×19 beam. This value comes to

$$\tfrac{3}{8} \text{ in.} + 0.30 \text{ in.} = 0.675 \text{ in.}$$

Next, the bottom cope must be measured from the bottom of the joist flange up to the top of the beam radius line. Since the W10×15 joist is $\tfrac{1}{4}$ in. less in depth than the W10×19 in. beam, it will require $\tfrac{1}{4}$ in. less depth in its cope on the bottom side, making the depth of the bottom cope

$$0.675 \text{ in.} - \tfrac{1}{4} \text{ in.} = 0.425 \text{ in.}$$

The length of the cope is equal to the a value of the W10×19 beam, which is $1\tfrac{7}{8}$ in. The cope radii need to equal the main beam radii of 0.30 in. This is all of the information required to detail the joist copes.

Our next step is to determine the size of rivets needed to fasten the joists, beams, and columns together. First, let us find the size of rivets needed for the joists. From Sample Exercise 7.1 we know that each joist carries a load of 6000 lb. This makes the value of S for one end of the joist $\tfrac{1}{2}(6000 \text{ lb}) = 3000$ lb. The 13-lb clip being used has two holes on each side of the mount side for a total of four rivets subjected to single shear. The size rivets for the mount side of the joists will be

$$d = \sqrt{\frac{3000 \text{ lb}}{2500\pi(4 \text{ rivets})}}$$

$$= 0.31 \text{ in. or } \tfrac{3}{8}\text{-in. rivets}$$

The web side of a 13-lb clip has four double-shear rivets making the required rivet size 0.31 in. × 1.41 = 0.437 in. or $\tfrac{1}{2}$-in. rivets.

Next, we'll size the main beam rivets. Since there are four main beams, each beam will also carry 6000 lb. This means that the sizing for the beam rivets will be exactly the same as that for the joist rivets. This will occur only when there are the same number of beams as joists.

Finally, we are ready to assemble the steel beam detail for this exercise. The result is shown in Figure 7.18. In this figure all of the information for the three different types of beams has been included on one drawing. Please study this detail carefully.

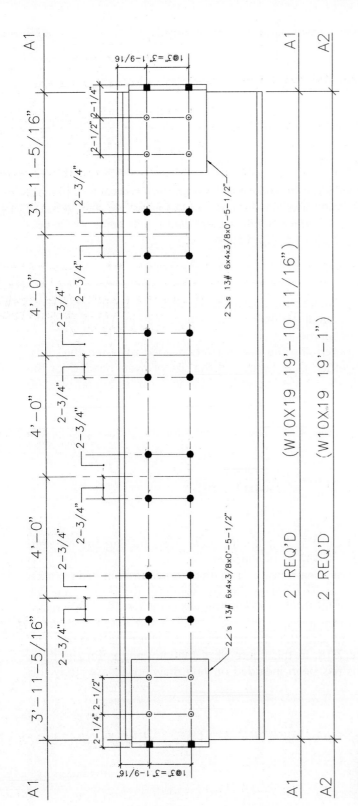

FIGURE 7.18

Problems

1. A steel beam spans two supports and is uniformly loaded. The beam is fastened at either end to the supports by a single steel bolt in single shear. Compute the size $\frac{1}{8}$-in. increment bolt needed to fasten this beam if it is loaded with

    (a) 10,000 lb
    (b) 5000 lb
    (c) 20,000 lb

2. Find the size $\frac{1}{16}$-in. increment bolts needed in Problem 1 if the bolts are in double shear.

3. Determine the size of the bolts needed to fasten the beam in Problem 1(a) if

    (a) 2 bolts are used at each end
    (b) 5 bolts are used at each end

4. Find the size bolts needed in Problem 3 if the bolts are in double shear.

5. Assuming standard steel clip orientation, determine the size steel rivets needed to fasten a beam carrying a 20,000-lb load to its supports at either end using

    (a) 13-lb clips
    (b) 19-lb clips
    (c) 33-lb clips

6. Compute the size WT type joists needed for each section of the structure in Problem 11 at the end of Chapter 6 if 10 joists are used per section.

7. Determine all of the clip and rivet sizes to fasten the structure in Problem 6. Assume that the main beams are all W12×50.

8. Detail any main beam and joist from Problem 6. Include all details and dimensions. Record any general notes as required.

CHAPTER 8

Wood Joist and Column Design and Detailing

8.1 INTRODUCTION

In this chapter we investigate the following topics as they relate to the selection, sizing, and detailing of structural wood members:

1. Moment of inertia of wood shapes
2. Wood nomenclature
3. Joist sizing for
    (a) Single members
    (b) Multiple members
4. Design and detailing for
    (a) Flooring
    (b) Decking
5. Wood column sizing

8.2 COMPUTATION OF MOMENT OF INERTIA

The term moment of inertia was introduced in Chapter 6 and it was shown that its value corresponds directly to the strength of a beam based on the shape and orientation of its cross section. In this section we examine the procedure used to compute the actual moment of inertia value for three specific shapes: rectangles, triangles, and circles. We also discuss the method used for finding moments of inertia for composite shapes.

Let us define the term moment of inertia as a numerical value that is based on the area and dimensions of a cross section. We will soon see that the larger this value is for a given shape, the stronger that shape will be.

First of all, it is important to understand why we need the value for moment of inertia in the first place. In Figure 8.1 we see three geometric shapes: a rectangle, a triangle, and a circle. The dimension b on each shape refers to the breadth of that cross section, which is the dimension parallel to the ground. The dimension d is the depth, or the dimension perpendicular to the ground. By computing various values of moment of inertia I we can determine the strongest orientation of the beam so that it can carry the greatest load. Using this value in conjunction with the standard deflection formula discussed in Chapter 6, we can determine maximum loads, beam and joist sizes, and maximum spans for structural members.

The axes shown on each shape pass through the centroid of the shape. The X-X axis on the rectangle runs parallel to the ground and the Y-Y axis runs perpendicular to the X-X axis. By computing I about a given axis we can determine how well a beam shape will withstand a load applied perpendicular to that axis.

FIGURE 8.1

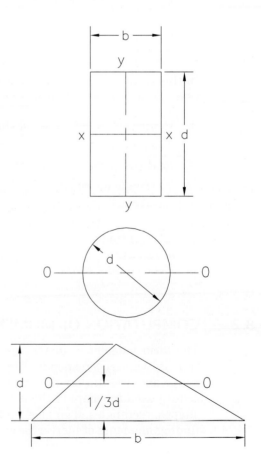

Rectangular Shapes

To begin this study, let us look at a wood floor joist that is 10 in. deep and 2 in. thick. When it is oriented as shown in Figure 8.2 we find that $b = 2$ in. and $d = 10$ in. The formula to compute moment of inertia for a rectangular shape is

$$I_{X\text{-}X} = \frac{bd^3}{12}$$

To find the value for $I_{X\text{-}X}$ for this 2-in.×10-in. joist, we use

$$I_{X\text{-}X} = \frac{(2 \text{ in.})(10 \text{ in.})^3}{12}$$
$$= 166.67 \text{ in}^4$$

If we want to know how this compares with the value of I measured against the Y-Y axis ($I_{Y\text{-}Y}$) we would simply swap the values for b and d:

$$I_{Y\text{-}Y} = \frac{(10 \text{ in.})(2 \text{ in.})^3}{12}$$
$$= 6.67 \text{ in}^4$$

So we see that by orienting the joist so that its largest dimension is parallel to the load applied we can increase it strength almost 25 times! This illustrates why orientation is so important when designing and locating structural members.

Triangular Shapes

Since triangles can be oriented in a variety of fashions we will concern ourselves only with a single orientation for any given problem. In most cases the base of the triangle will be perpendicular to the line of force, which for a beam will mean that the base is parallel to the ground. The axis against which the force(s) upon the beam will be acting runs through the centroid, which is located one-third of the way from the base along its altitude. We refer to this axis as the zero (0) axis because it is the line running through the origin point of the shape. The moment of inertia about an axis through a triangle is

$$I_0 = \frac{bd^3}{36}$$

FIGURE 8.2

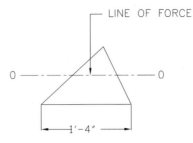

FIGURE 8.3

This value for the triangular beam shown in Figure 8.3 is

$$I_0 = \frac{(16 \text{ in.})(10 \text{ in.})^3}{36}$$
$$= 444.44 \text{ in}^4$$

The strongest orientation for a triangular-shaped member can be found by computing the value for I_0 for each of its three orientations and comparing them to find the largest result.

Circular Shapes

Now let us look at perhaps the easiest shape to understand, the circle. No matter how you turn it, a circle always will have the same dimension for depth, which is its diameter (see Figure 8.4). The formula for moment of inertia of a circle is

$$I_0 = \frac{\pi d^4}{64}$$

Therefore, the moment of inertia for a round beam 2 in. in diameter would be

$$I_0 = \frac{\pi (2 \text{ in.})^4}{64}$$
$$= 0.785 \text{ in}^4$$

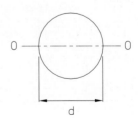

FIGURE 8.4

 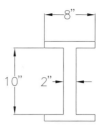

FIGURE 8.5

Composite Shapes

Of course, not all shapes are simple, solid geometric figures. Some are much more complex. Many of the more complex shapes, such as the I-beam, are actually made up of several simple shapes.

Let us take as our example the I-beam section shown in Figure 8.5. An I-beam is actually a rectangular beam section with two rectangular sections removed from either side. We can compute the moment of inertia of such a shape by first finding the moment of inertia of the entire shape and then subtracting out the moments of inertia of the rectangular spaces. Using the formula for rectangular shapes we find

$$I_{X\text{-}X(\text{RECT})} = \frac{(8 \text{ in.})(12 \text{ in.})^3}{12}$$
$$= 1152 \text{ in}^4$$

This is the moment of inertia for the entire 8-in. × 12-in. rectangle. Within that rectangle, however, there are two rectangular spaces that need to be removed. Each rectangle is 10 in. high by 3 in. wide. We can put them

This deck is a pine wood structure consisting of 6 joists with 2 end joists and 2 header joists. The two square posts in front are designed to support one-half the total deck load (the back side of the deck is mounted to the exterior wall of the house). The surface of this deck in made of pine deck boards. This is an excellent example of the use of structural wood members to construct a load-bearing wood structure.

FIGURE 8.6

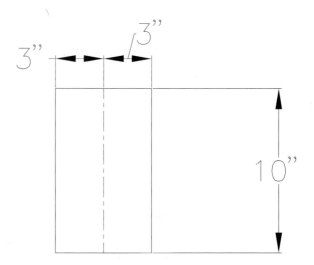

together to make one 10-in. × 6-in. rectangle, as shown in Figure 8.6, and find its composite moment of inertia. The result will be

$$I_{X\text{-}X(\text{SPACE})} = \frac{(3 \text{ in.} + 3 \text{ in.})(10 \text{ in.})^3}{12}$$

$$= 500 \text{ in}^4$$

$$\therefore I_{X\text{-}X(\text{I-BEAM})} = 1152 \text{ in}^4 - 500 \text{ in}^4 = 652 \text{ in}^4$$

This principle holds true for all composite shapes as well. The moment of inertia of a composite shape is equal to the sum of moments of inertia of all the individual shapes or the difference between moments of inertia of overall shapes and the spaces inside them. Later in this chapter we will see how the value for moment of inertia of a wood joist is used in sizing, load-bearing, and length computations.

8.3 WOOD NOMENCLATURE

To properly do the work required to design and detail structural wood members it is important for you to understand some basic terms and definitions about wood. In this section we discuss these items as they relate specifically to a type of pine wood known as Douglas fir, a wood commonly used for structural members in residential construction.

Let us first look at some definitions of wood members and wood structures as shown in Figure 8.7:

Decking—The surface of the flooring that rests upon the joists.

Flooring—The materials used to construct a floor. Consists mainly of joists and decking.

Joist—Horizontal structural support member spaced at regular intervals under flooring or decking.

FIGURE 8.7

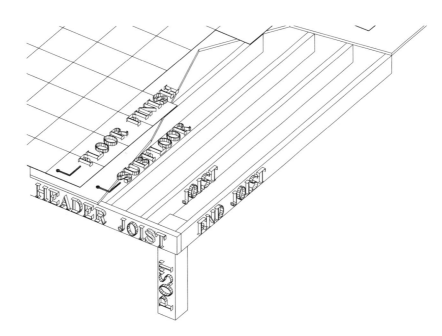

End joists—Joists that run outside of the floor frame parallel to the floor joists.

Header joists—Joists that run outside of the floor frame perpendicular to the floor joists.

Post—Also called a column. Vertical structural support member.

Solid sawn—A member that has been cut and shaped from a single piece of wood.

Subflooring—The part of the decking that rests directly on the joists. Does not include carpet or other floor finishes.

Wood sizes are referred to in one of two ways: nominal or actual. **Nominal** wood sizes are the named sizes. In most cases these sizes are not the true dimensions of the wood. **Actual** wood sizes are the true dimensions of a piece of wood after it has been finished to specifications. In the performance of design computations the actual dimensions are used. **Milling** specifications are used to determine the actual size of a piece of lumber. When the designer is in doubt of the actual milling value for a particular dimension then the milling value that yields the smallest dimension, or **minimum material condition** (also called least material condition), should be assumed. Nominal and actual dimensions refer specifically to the depth and breadth dimensions of a piece of lumber and not to its length.

Figure 8.8 shows the cross section of a nominal 2×4 stud. This stud has been milled $\frac{3}{8}$ in. along its breadth and $\frac{1}{2}$ in. along its depth. Although we call it a 2×4, it is actually $1\frac{5}{8}$ in.×$3\frac{1}{2}$ in. It would be awkward to call it a "one and five-eighths by three and a half" though, so we give it a nominal designation, a name—we call it a "two-by-four." We understand that it is not

FIGURE 8.8

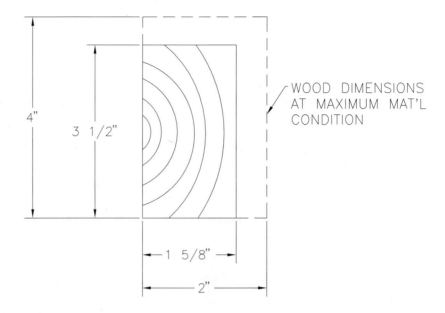

exactly two inches by four inches, but was milled to a given set of specifications from a 2-in. by 4-in. piece of lumber. We only assume that the actual and nominal dimensions are the same when no other information is available. Even then, it is wise to include the note, "Nominal dimensions treated as actual for computations."

To illustrate the importance of using the procedure we have just discussed let us consider the following example. Compute the moment of inertia for a 2×8 floor joist assuming maximum allowed milling values of $\frac{1}{2}$ in. along its depth and $\frac{1}{8}$ in. along its breadth. Using the milling values given we find the actual dimensions to be

$$\text{Breadth} = 2 \text{ in.} - \tfrac{1}{8} \text{ in.} = 1\tfrac{7}{8} \text{ in.}$$
$$\text{Depth} = 8 \text{ in.} - \tfrac{1}{2} \text{ in.} = 7\tfrac{1}{2} \text{ in.}$$

Since the cross section of a 2×8 joist is a rectangle, we use the following to find its moment of inertia:

$$I_{X\text{-}X} = \frac{(1\tfrac{7}{8} \text{ in.})(7\tfrac{1}{2} \text{ in.})^3}{12}$$
$$= 65.92 \text{ in}^4$$

If we had failed to account for the milling in this problem and had used the nominal values for breadth and depth, our calculations would have falsely revealed that $I_{X\text{-}X} = 85.33$ in^4. In reality this error would have made this joist appear almost 30% stronger than it actually is—a potentially disastrous error when designing habitable structures! Furthermore, errors compound each other for every extra floor added to a structure. A three-story structure containing the error shown in the example above would have a

41% error in its ground floor members! This makes it easy to see why actual dimensioning is so important.

Just like steel (or any other substance in the universe), wood has weight, but unlike steel, wood member weights are not generally indicated in pounds per linear foot. Instead, a unit called the board-foot is used. One **board-foot** of wood is based on the dimensions of a 1-in.-thick by 12-in.-wide piece of lumber that is 1 ft in length. To compute the number of board-feet in a piece of lumber we use the following formula:

$$\text{bf} = \frac{t(w)l}{12}$$

where

bf = number of board-feet
t = thickness of lumber in inches
w = width of lumber in inches
l = length of lumber in feet

Douglas fir weighs 2.65 lb/bf. To find the weight of a length of lumber we compute the number of board-feet and multiply the result by its weight per board-foot. For example, if a floor surface was made up of twenty 12-ft-long 1-in. × 12-in. finished boards, its total weight would be

$$\text{bf} = \frac{1 \text{ in.}(12 \text{ in.})(12 \text{ ft})}{12}$$

bf (per board) = 12 bf
total board-feet = 20 boards × 12 bf/board = 240 bf
total weight = 2.65 lb/bf × 240 bf = 636 lb

An understanding of the terms discussed in this section is necessary when performing structural wood member computations. In some ways, wood calculations are easier than steel calculations, but sometimes the number of variables and dimensions can become a bit overwhelming. For this reason it is important for you to keep your work well organized. In the event that you do make a mistake it will be much easier to find and correct.

Now that we have covered an adequate amount of background information on wood, we are ready to begin our detailed study of structural wood design.

8.4 SIZING SINGLE WOOD MEMBERS

To determine the size of a single wood member needed to support a given load we will follow a procedure very similar to that shown in Section 6.8. There are, however, some differences in the calculations for lumber, as we shall soon see.

First of all, most structural wood members used to support a floor load have even-numbered dimensions. For example, 2×8, 2×10, and 4×6 are common joist sizes, while 3×6 and 3×5 are not standard wood sizes. This is important to remember because you will now be required to calculate the actual size of the joist instead of finding it in a chart. Also, the values for modulus of elasticity E vary, depending on the type and grade of wood used. To find this value for the problems listed in this text you will need to refer to the Douglas fir data in Table C in the back of the book. The standard allowable live-load deflection value for structural lumber is $\frac{1}{240}$ of the total span. The length of the span is measured in inches and the unbraced horizontal dimension is used. Using these specifications let us now look at a sample problem.

Compute the size of a grade 2 Douglas fir joist needed to carry a live-load of 1000 lb along an 8-ft span.

1. Write the necessary formula:

$$I_{X\text{-}X} = \frac{5Wl^3}{384ED}$$

2. List all given information:

   W = 1000 lb
   l = 96 in.
   E = 1,600,000 psi (for grade 2 Douglas fir, found in Table C in the back of this text)

3. Compute the allowable deflection:

$$\delta = \Delta y / x \quad \delta = \tfrac{1}{240}, \quad x = 96 \text{ in.}$$
$$D = \Delta y = \delta(x) = \tfrac{1}{240} \times 96 \text{ in.} = 0.4 \text{ in.}$$

4. Substitute all known values into the $I_{X\text{-}X}$ formula and compute the required value for $I_{X\text{-}X}$:

$$I_{X\text{-}X} = \frac{5(1000 \text{ lb})(96 \text{ in.})^3}{384(1{,}6000{,}000)(0.4 \text{ in.})}$$
$$= 18 \text{ in}^4$$

5. Compute the actual dimensions of the joist. We know that the formula for $I_{X\text{-}X}$ requires that we know both the breadth and depth of a rectangular shape. Since we don't currently know either, we must assume one of the values. If the space in which our joist is to be located limits its depth, then we'll assume that limit as the depth of our joist. If no such limit exists, then we can assume a value for the breadth of the joist.

   Since we know that standard dimensions of structural lumber are usually measured in increments of 2 in. we will assume 2 in. as our joist breadth. Then, rewriting the formula for moment of inertia of rectangular shapes, we find

$$d = 3\sqrt{\frac{12 I_{X\text{-}X}}{b}}$$

Assuming a breadth of 2 in. we find

$$d = 3\sqrt{\frac{12(18)}{2 \text{ in.}}}$$
$$= 4.76 \text{ in.}$$

6. Record the nominal joist to be used. Sizing the depth dimension of 4.76 in. up to the next even number, we find that a nominal 2×6 grade 2 Douglas fir joist would be an appropriate selection for this member. Since the actual area of the cross section required is

$$A = 4.76 \text{ in.} \times 2 \text{ in.} = 9.52 \text{ in}^2$$

the 2×6 selected can be milled as necessary as long as the resulting depth is not less than 4.76 in. and the overall area of the cross section is not less than 9.52 in² at minimum material condition.

An example of how you might detail and clarify the information found in the previous sample problem is shown in Figure 8.9. The notes and dimensions shown correspond to the actual conditions that must be maintained for this joist to support the required load. Notice that the required values for cross-section area and depth have been increased. This creates a built-in safety factor so that the joist is guaranteed to do the job for which it has been designed.

While we are looking at this 2×6 joist, let us further investigate this member under specific field requirements. Let's assume that we've been told that this joist must be placed under a ¾-in. plywood deck so that the total

FIGURE 8.9

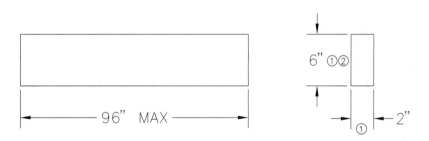

NOMINAL 2X6 DETAIL

NOTES:
① DIMENSIONS MAY BE MILLED AS NECESSARY TO MAINTAIN CROSS-SECTION AREA OF 10 IN SQ. OR GREATER, SUBJECT TO NOTE ② CONDITION.
② DIMENSION MUST REMAIN 5 IN. OR GREATER AT MIN. MATERIAL CONDITION.
③ SPECIFY ACTUAL DIMENSIONS AS PER FIELD REQUIREMENTS.

FIGURE 8.10

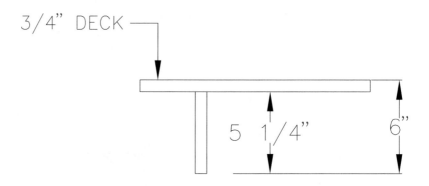

depth of the deck and joist is 6 in. This means that the joist must be milled to a depth of $5\frac{1}{4}$ in. (see Figure 8.10). We can then determine the maximum milling value for the breadth of the joist by

$$10 \text{ in}^2 = 5\tfrac{1}{4} \text{ in.} \times b_{\text{MIN}}$$
$$\therefore b_{\text{MIN}} = 1.90 \text{ in. or } 1\tfrac{7}{8} \text{ in.}$$

So, the actual dimensions of this joist would be $5\frac{1}{4}$ in. $\times$ $1\frac{7}{8}$ in. and its actual $I_{X\text{-}X} = 22.6$ in^4, which allows for a 25% safety margin over the required $I_{X\text{-}X}$ value of 18 in^4.

Existing Structures or Materials

If a structure has already been built it may be necessary to determine the maximum load capacity of that structure. For instance, we might have a post-and-beam support in the basement of a house with a 4×8 select structural Douglas fir beam resting on posts located 8 ft O.C. Assuming the beam is a true 4-in. × 8-in. beam, we will use the following procedure to compute the allowable load on an 8-ft section of the support structure.

1. Write the appropriate formula:

$$W_{\text{MAX}} = \frac{384 E I_{X\text{-}X} D}{5 l^3}$$

2. List all known information:

$$E = 1{,}900{,}000 \text{ psi} \quad (\text{found in Table C})$$
$$l = 96 \text{ in.}$$

2a. Compute $I_{X\text{-}X}$:

$$I_{X\text{-}X} = \frac{(4 \text{ in.})(8 \text{ in.})^3}{12}$$
$$= 170.67 \text{ in}^4$$

3. Compute the allowable live-load deflection:

$$D = \Delta y = \delta(x) = \tfrac{1}{240} \times 96 \text{ in.} = 0.4 \text{ in.}$$

4. Compute the maximum allowable live load on the beam:

$$W_{MAX} = \frac{384(1{,}900{,}000)(170.67)(0.4 \text{ in.})}{5(96 \text{ in.})^3}$$
$$= 11{,}259.48 \text{ lb}$$

The value of 11,259.48 lb represents the maximum load that can be placed on a single 8-ft section of this post-and-beam support. If the basement is 24 ft long, then the total live load on the floors above could be about 30,000 lb with about a 10% safety margin.

We can also compute the maximum length of a span if we know the size and type of joist and the maximum live load that it must withstand. Rewriting the standard deflection formula to solve for length, we have

$$l = 3\sqrt{\frac{384EID}{5W}}$$

Let us use as our example the log in Figure 8.11. If the log is to be used as a bridge to transport a 2-ton load of rocks across a ravine and it has a modulus of elasticity equal to a Douglas fir grade 3 beam, how wide can the ravine be if the log becomes unstable at 1 in. of deflection?

1. Write the appropriate formula:

$$l_{MAX} = 3\sqrt{\frac{384EID}{5W}}$$

2. List all known information:

$E = 1{,}400{,}000$ psi (from Table C for Douglas fir grade 3)
$W = 4000$ lb
$D = 1$ in.

FIGURE 8.11

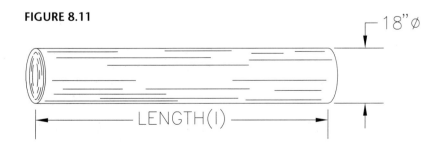

Structural wood members, such as these roof trusses, are often evenly spaced over an area to support a load. The bases of these roof trusses act very much like joists in carrying the live and dead roof loads. *Courtesy of Krieser Construction.*

2a. Calculate the moment of inertia for the shape. Since the log is round we use

$$I_0 = \frac{\pi d^4}{64}$$
$$= \frac{\pi (18 \text{ in.})^4}{64}$$
$$= 5150.39 \text{ in}^4$$

3. Solve for the maximum allowable length of the span:

$$l_{MAX} = 3\frac{384(1,400,000)(5150.39)(1 \text{ in.})}{5(4000 \text{ lb})}$$
$$= 517.32 \text{ in. or about 43 ft}$$

So the log could span a ravine up to about 40 ft wide with little risk of failing under the 2-ton load.

Now that we have seen how to size individual wood members, let us move on to look at some conditions in which several wood members are used to carry a load.

8.5 SIZING MULTIPLE WOOD MEMBERS

The procedure for sizing multiple wood members such as beams or joists is the same as the one used in Chapter 7 for sizing multiple steel supports. Let us look at an example of a section of flooring 10 ft wide by 15 ft long that rests on unmilled 2×6 select structural Douglas fir joists spaced 1 ft O.C. spanning the longest dimension of the floor (see Figure 8.12). To determine the maximum live load capacity of this floor section we will use the following procedure

FIGURE 8.12

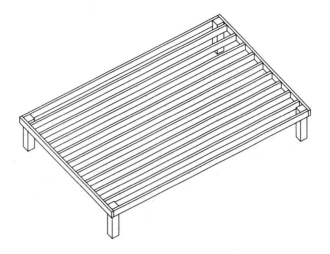

1. Write the formula for maximum load capacity:

$$W_{MAX} = \frac{384 E I_{X\text{-}X} D}{5 l^3}$$

2. Record all known information:

   $E = 1,900,000$ psi (for select structural from Table C)
   $l = 15$ ft $\times$ 12 in./ft $= 180$ in. (longest dimension)

3. Compute remaining values from information given.

   (a) Moment of inertia—Since the joists are unmilled we can assume the actual size is the same as the nominal size:

   $$I_{X\text{-}X} = \frac{(2 \text{ in.})(6 \text{ in.})^3}{12}$$
   $$= 36 \text{ in}^4$$

   (b) Deflection:

   $$D = D = \Delta y = \delta(x) = \tfrac{1}{240} \times (15 \text{ ft} \times 12 \text{ in./ft}) = 0.75 \text{ in.}$$

   (c) Number of joists:

   $$j = \frac{w_{\text{FLOOR}}}{\text{spacing}} + 1$$

where

   j = number of joists
   w_{FLOOR} = floor width $\perp$ joists in inches
   spacing = O.C. joist spacing in inches

$$j = \frac{120 \text{ in.}}{12 \text{ in.}} + 1$$
$$= 11 \text{ joists}$$

4. Compute the maximum load capacity for a single joist:

$$W_{MAX} = \frac{384 E I_{X-X} D}{5 l^3}$$
$$= \frac{384(1{,}900{,}000)(36)(0.75)}{5(180)^3}$$
$$= 675.56 \text{ lb per joist}$$

5. Compute the maximum load capacity for the floor section. Since there are 11 joists, then

$$W_{MAX(SECTION)} = j(W_{MAX}) = 675.56 \text{ lb per joist} \times 11 \text{ joists}$$
$$= 7431 \text{ lb}$$

It would, as always, be wise to include a safety factor in your design work. In the case of the problem we have just examined it would be a good idea to allow about a 10% safety margin, which would yield about a 6600- to 6700-lb load capacity.

Remember, when performing the computations for any flooring section that is supported by more than one beam or joist you will be solving for the missing information for only a single wood member. Remember always to account for all of the support members by the end of the problem. Failure to do so will result in gross oversizing or undersizing of the structural elements.

Now we'll look at some examples of calulation involving real floors and decks.

With the correct design wood members can span great distances. As you can see in this photograph there is a great deal of open space on the first floor of this house because no columns or posts are needed to support the roof load from inside this space. *Courtesy of Krieser Construction.*

8.6 DECK AND FLOOR SECTION DESIGN

Figure 8.13 shows a length of deck made of Douglas fir. The decking boards are 2×12's and the joists are grade 1 2×8's. Using this scenario we will recommend a maximum safe length for this section of the deck if it must carry a minimum live load of 5200 lb. For this problem we must keep in mind that the decking material, in this case the 2×12 boards, will add to the overall weight of the structure. This will force us to include an additional step at the end of this problem. First, however, let us solve this problem using the technique with which we are familiar.

1. Write the appropriate formula:

$$l_{MAX} = 3\sqrt{\frac{384EID}{5W}}$$

We must pause at this point because without knowing the actual length of the deck section, we do not know the actual length of the joists. Since we don't know their length we can't compute their deflection value. To compensate for this situation we simply rewrite the deflection formula, substituting $\delta(l)$ for D ($l = x$ in this substitution), and generate the following formula:

$$\delta(l) = \frac{5Wl^3}{384EI_{X\text{-}X}}$$

Solving for l we find

$$l_{MAX} = \sqrt{\frac{384EI_{X\text{-}X}\delta}{5W}}$$

FIGURE 8.13

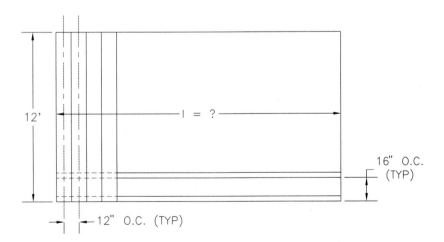

This formula allows us to compute the maximum allowable length of a member based on its allowable deflection ratio δ, since we cannot compute its actual deflection in inches.

2. List all known information:

$$E = 1{,}600{,}000 \text{ psi} \quad \text{(from Table C for Douglas fir grade 1)}$$
$$\delta = \tfrac{1}{240}$$

2a. Calculate the moment of inertia for the shape. Since the joists are rectangular we use

$$I_{X\text{-}X} = \frac{bd^3}{12}$$

Without any milling information we will assume that the actual size of the joists is the nominal size.

$$I_{X\text{-}X} = \frac{2 \text{ in.}(8 \text{ in.})^3}{12}$$
$$= 85.33 \text{ in}^4$$

2b. Compute the maximum live load per joist. First we find the number of joists by

$$j = \frac{w_{\text{FLOOR}}}{\text{spacing}} + 1$$

From Figure 8.13 we see

$$W_{\text{FLOOR}} = 12 \text{ ft} = 144 \text{ in.}$$
$$\text{spacing} = 16 \text{ in.}$$
$$j = \frac{144 \text{ in.}}{16 \text{ in.}} + 1$$
$$= 10 \text{ joists}$$

Then, we find the load per joist by

$$W = 5200 \text{ lb}/10 \text{ joists} = 520 \text{ lb/joist}$$

3. Solve for the maximum allowable length of the span:

$$l_{\text{MAX}} = \sqrt{\frac{384(1{,}600{,}000)(85.33)(\tfrac{1}{240})}{5(520 \text{ lb})}}$$
$$= 289.86 \text{ in. or about } 24 \text{ ft}$$

Since lumber lengths are most commonly cut in increments of 2 ft, this length would be a good choice.

4. There is one final item to consider that was mentioned at the beginning of this problem. To be safe we will treat the weight of the decking boards as live loads in order to prevent excessive deflection. This will make the total weight on this section 5200 lb plus the decking weight.

$$288 \text{ in.} = \text{deck length}$$
$$12 \text{ in.} = \text{deck board spacing}$$
$$bf_{\text{(DECKING)}} = \left(\frac{2 \text{ in.} \times 12 \text{ in.} \times 12 \text{ ft}}{12}\right)\left(\frac{288 \text{ in.}}{12 \text{ in.}}\right)$$
$$= 576 \text{ bf}$$
$$= 576 \text{ bf} \times 2.65 \text{ lb/bf} = 1526.4 \text{ lb}$$

So the total weight of the deck would be 6726.4 lb, making the load per joist 672.6 lb.

The allowable deflection for a 240-in. span of lumber is 1 in., so the deflection value of this span must not exceed 1 in. (because $\frac{1}{240}$ of 240 in. is 1 in.). To test this condition we use the standard deflection formula:

$$D = \frac{5Wl^3}{384EI_{X\text{-}X}}$$

$$= \frac{5(672.6 \text{ lb})(288 \text{ in.})^3}{384(1{,}600{,}000)(85.3)}$$

$$= 0.887 \text{ in.}$$

Since 0.887 in. < 1 in. the length of 24 in. will be safe to use for one deck section. If the computed value of D had been greater than the allowed value of 1 in., the deck length would need to be reevaluated in the same manner shown in step 4, using a shorter length of section. We would continue proceeding in this manner until the computed deflection was less than the allowed deflection.

Note: You will need to recalculate the allowed deflection δl each time you resize the length of the section. Failure to do so will result in a false comparison between the allowed and actual deflection values.

The most important thing to remember when performing computations on multiple load-bearing members is that you must always work with one member only. This generally means that you will be dividing the load on a section by the number of load-bearing members that it contains. It is also important to recognize that the procedures discussed in this chapter are all based on uniformly distributed loads placed on evenly spaced members.

Armed with the information discussed thus far in this chapter we are now ready to take a look at the method used to size vertical support members.

The multiple wood studs used to frame a wall, such as the interior wall shown in this photograph, all act as individual wood posts. Their strength is combined in supporting the loads placed on the walls. Notice the four studs sandwiched together near the center of the picture. They act as a single post upon which a load-bearing portion of the roof rests. *Courtesy of Krieser Construction.*

8.7 SOLID SAWN WOOD COLUMNS

Wood members are significantly stronger under compression than they are under tension or shear. For this reason, as you will see, it does not take a very large wood column to carry a large amount of weight. When we size a wood column, or any column for that matter, we do so for the **total load** that column will carry, not just the live load. The procedure for sizing wood columns is very simple, but the formula is quite cumbersome. For this reason, we explain in detail each of the elements required for the solution of wood columns, and then offer a method to simplify the **column compression equation**.

The equation used to solve for the maximum allowable compression on a wood column is:

$$P = F_c A \left\{ \frac{1 + K_{ce}Ed^2/F_c L_e^2}{2c} - \sqrt{\left[\frac{(1 + K_{ce}Ed^2)/F_c L_e^2}{2c}\right]^2 - \frac{K_{ce}Ed^2/F_c L_e^2}{c}} \right\}$$

where

P = compression on wood column in pounds

F_c = allowable stress on column parallel to grain in psi (Table C)

A = area of column cross section in in^2

K_{ce} = 0.3 for visually graded lumber, 0.418 for glue-laminated lumber

Visually graded lumber—Hand-finished lumber without the use of fine machine measurements.

Glue-laminated lumber—Wood laminated with an epoxy. Stronger and smoother than hand-finished lumber.

E = modulus of elasticity in psi (Table C)
d = column width in direction of buckling in inches (usually the shortest dimension for all non-square or round columns)
L_e = unbraced length of the column in inches
c = 0.8 for sawn lumber, 0.85 for round lumber, 0.9 for glue-laminated

Sawn lumber—Cut and unfinished.
Round lumber—Cut into a round cross section, usually using a lathe.
Glue-laminated lumber—See K_{ce} above.

This equation seems quite intimidating at first, but if you look at it closely a pattern will appear. The expression

$$\frac{K_{ce} E d^2}{F_c L_e^2}$$

appears three times in this equation. If we substitute another single variable for it, say Q, then we only have to solve for this expression one time. We can then substitute the value found for Q into the equation below:

$$P = F_c A \left[\frac{1 + Q}{2c} - \sqrt{\left(\frac{1 + Q}{2c}\right)^2 - \frac{Q}{c}} \right]$$

Let us look at a problem that involves the use of this formula for sizing wood columns. For our example we will again use the deck section shown in Figure 8.13. If the section is to be supported by 10-ft-tall 4-in.×4-in. unfinished Douglas fir grade 2 posts, will these posts be sufficiently strong to support the deck?

We'll assume that the load is evenly distributed among the four posts so that the value we find will actually represent one-fourth of the load capacity for the entire deck section. Given that the total load on this section is 6800 lb (rounded up from the actual value of 6726.4 lb) plus the weight of the ten 2×8 joists (24 ft long each = 32 bf/joist), we will be testing to see if the 4-in.×4-in. post is strong enough to support a load of $\frac{1}{4}$(6800 lb + 848 lb), or 1912 lb. Let us also assume that it would take an additional 20 lb of fasteners to hold this assembly together. This brings the total load to 1932 lb, so let's round up to 1940 lb. To find the answer we use the following procedure

1. Write the appropriate formulas:

$$P = F_c A \left[\frac{1 + Q}{2c} - \sqrt{\left(\frac{1 + Q}{2c}\right)^2 - \frac{Q}{c}} \right]$$

$$Q = \frac{K_{ce} E d^2}{F_c L_e^2}$$

2. List all information needed to solve for Q:

$K_{ce} = 0.3$ (unfinished lumber is visually graded)
$E = 1,600,000$ psi (Table C)
$d = 4$ in.
$F_c = 1300$ psi (Table C value parallel to grain)
$L_e = 10$ ft $= 120$ in.

2a. Solve for Q:

$$Q = \frac{(0.3)(1,600,000)(4 \text{ in.})^2}{(1300)(120 \text{ in.})^2}$$
$$= 0.410$$

2b. List all remaining information need to solve for P:

$A = 4$ in. $\times$ 4 in. $= 16$ in^2
$c = 0.8$ (unfinished lumber is sawn, not finished or laminated)

3. Solve for P:

$$P = (1300 \text{ lb})(16) \left\{ \frac{1 + (0.410)}{2(0.8)} - \sqrt{\left[\frac{1 + (0.410)}{2(0.8)}\right]^2 - \frac{0.410}{0.8}} \right\}$$
$$= (20,800)\left(0.881 - \sqrt{(0.881)^2 - 0.513}\right)$$
$$= 7654.4 \text{ lb}$$

This post can hold 7654.4 lb, which is greater than the actual load of 1940 lb that it must carry. Obviously, this post is more than strong enough to perform the job for which it has been chosen.

Extra framing around door and window rough openings provides added stability against the loads applied to these areas. Notice the double studding on either side of the door opening and the header at the top of the opening. Each of these structural elements increases the structural integrity of the opening. *Courtesy of Krieser Construction.*

Estimating Wood Column Sizes

It is possible to compute the size of a given wood column to carry a known load if we assume the dimension of the column in the direction of buckling (d). We can compute the necessary area of the column and, from that dimension, derive the appropriate rectangular shape.

Let us look at the column from the previous problem. We know that it must carry a load of 1860 lb and we have already seen that a 4-in² column is more than sufficient for this purpose. Let us, then, assume a smaller dimension in the direction of buckling. We'll use $d = 3$ in. for this problem.

1. If we rewrite the column compression formula to solve for the area A of cross section, we have

$$A = \frac{P}{F_c\left[\dfrac{1+Q}{2c} - \sqrt{\left(\dfrac{1+Q}{2c}\right)^2 - \dfrac{Q}{c}}\right]}$$

$$Q = \frac{K_{ce} E d^2}{F_c L_e^2}$$

2. Solving for Q we find

$$Q = \frac{0.3(1{,}600{,}000)(3 \text{ in.})^2}{(1300)(120)^2}$$

$$= 0.231$$

3. With $P = 1860$ lb (required capacity of one post) we solve for A:

$$A = \frac{1860 \text{ lb}}{(1300)\left\{\dfrac{1+0.231}{2(0.8)} - \sqrt{\left[\dfrac{1+0.231}{2(0.8)}\right]^2 - \dfrac{0.231}{0.8}}\right\}}$$

$$= \frac{1860 \text{ lb}}{(1300)\left[0.769 - \sqrt{(0.769)^2 - 0.289}\right]}$$

$$= 6.52 \text{ in}^2$$

This tells us that with a 3-in. depth the area of cross section of this post must be 6.52 in². This would require the other dimension to be 6.52 in²/3 in. = 2.17 in. An actual 3×3 post would be adequate for this job.

Note: If the remaining dimension found in a rectangular post is greater than the assumed dimension d in the direction of buckling, then you must make the post square using the assumed dimension as the side length. If this results in gross oversizing, then you will need to assume a smaller value for d and recalculate the missing dimension.

For practice, see if a nominal 3×3 with ½-in. max. milling along each dimension would also work for this post. What method will you use to test this condition? Why?

8.8 SUMMARY

In this chapter we have seen some of the many uses of wood in the formation of a load-bearing structure. We now understand how wood dimensions are labeled and the difference between nominal and actual wood dimensions. Additionally, we have learned how to compute the number of board-feet in a given piece of lumber, and we have discovered how we can use this value to find the weight of the piece(s) of lumber in question. You now have a better understanding of some of the methods that can be used to size wood joists and beams to carry a give load. Also, you have learned to compute or estimate the dimensions for a wood post or column used to support a uniform load.

Now, let us consider some practical examples of wood structures in which the principles discussed in this chapter could be utilized.

Sample Exercise 8.1

Using Figure 8.14, compute the maximum weight of the structure that can be placed on this Douglas fir select structural grade post-and-beam assembly, assuming that the posts and walls are evenly spaced.

1. Write the appropriate formula:

$$W_{MAX} = \frac{384EI_{X-X}D}{5l^3}$$

2. List all known information:

$E = 1,900,000$ psi (found in Table C)
$l = (22 \text{ ft} \times 12 \text{ in./ft})/3 \text{ segments} = 88 \text{ in.}$

FIGURE 8.14

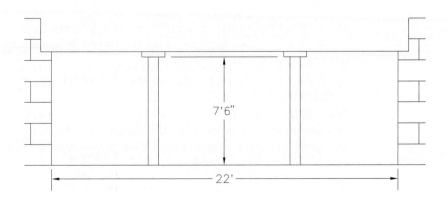

The base of the multiple member wood truss near the top of this photograph acts as a header joist for the bases of all of the other roof trusses. The last truss base on the top left of the photo acts as an end joist. *Courtesy of Krieser Construction.*

2a. Compute I_{x-x}:

$$I_{x-x} = \frac{(6 \text{ in.})(6 \text{ in.})^3}{12}$$
$$= 108 \text{ in}^4$$

3. Compute the allowable live-load deflection:

$$D = \Delta y = \delta(x) = \tfrac{1}{240} \times 88 \text{ in.} = 0.37 \text{ in.}$$

4. Compute the maximum allowable live load on the beam:

$$W_{MAX} = \frac{384(1{,}900{,}000)(108)(0.37 \text{ in.})}{5(88 \text{ in.})^3}$$
$$= 8556.42 \text{ lb per section}$$

Since the beam is made up of 3 evenly spaced sections, each capable of supporting 8556.42 lb, the total capacity of the system would be 3×8556.42 lb, or

$$W_{MAX} = 25{,}669.26 \text{ lb}$$

Allowing for a reasonable safety factor ($\approx 6.5\%$) we can say that a structure that has an internal weight of 12 tons could safely be placed on this post-and-beam system.

In Sample Exercise 8.3 we will look more closely at the requirements for the posts in this system.

Sample Exercise 8.2

The platform shown in Figure 8.15 must carry an additional live load of 20,000 lb. Treating the weight of the decking as a live load, compute the number and spacing of Douglas fir grade 1 or better joists needed to support this structure.

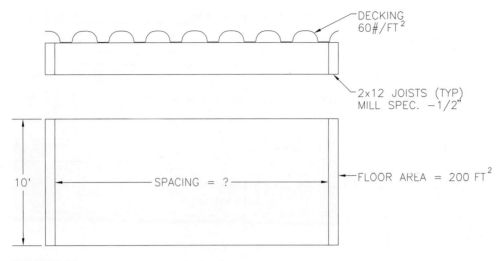

FIGURE 8.15

Since this problem does not match any of the scenarios we have discussed let us use the following procedure.

1. List all known information:

$$W_{SECTION} = 20{,}000 \text{ lb} + 200 \text{ ft}^2(60 \text{ psf}) = 32{,}000 \text{ lb}$$
$$l = 10 \text{ ft} = 120 \text{ in.}$$
$$D = \tfrac{1}{240} \times 120 \text{ in.} = 0.5 \text{ in.}$$
$$E = 1{,}800{,}000 \text{ psi}$$

2. Compute $I_{x\text{-}x}$ for one joist. Using the minimum material condition for the milling specifications in the diagram we let $b = 1\tfrac{1}{2}$ in. and $d = 11\tfrac{1}{2}$ in.:

$$I_{x\text{-}x} = \frac{1.5 \text{ in.}(11.5 \text{ in.})^3}{12}$$
$$= 190.11 \text{ in}^4$$

3. Find the load capacity for one joist:

$$W_{MAX} = \frac{384 E I_{x\text{-}x} D}{5 l^3}$$

$$= \frac{384(1{,}800{,}000)(190.11)(0.5)}{5(120)^3}$$

$$= 7604.4 \text{ lb per joist}$$

4. Compute the number and spacing of joists needed:

$$\text{Number of joists} = 32{,}000 \text{ lb}/7604.4 \text{ lb per joist}$$
$$= 4.2$$

Therefore, 5 joists are required. Rewriting the joist spacing formula we find

$$\text{Spacing} = \frac{W_{section}}{j - 1}$$

$$= \frac{20 \text{ ft}}{(5 - 1) \text{ joists}}$$

$$= 5 \text{ ft O.C.}$$

Sample Exercise 8.3

For our last example we'll look at a problem in which we are required to size a round wooden post in a post-and-beam system. Using the values from Sample Exercise 8.1 we will determine the size post required to support this structure.

1. First we must assume a value for d. This will automatically generate a value for A since the area of a circle is dependent on its diameter. If we assume a 3-in. post, then its area of cross section will be 7.065 in². If the value we compute for A using the column compression equation is less than 7.065 in², then the post diameter we have chosen will work. If not, then we will need to assume another size and rework the problem.

2. Next, we write the formulas necessary to solve this problem:

$$A = \frac{P}{F_c \left[\frac{1 + Q}{2c} - \sqrt{\left(\frac{1 + Q}{2c}\right)^2 - \frac{Q}{c}} \right]}$$

$$Q = \frac{K_{ce} E d^2}{F_c L_e^2}$$

3. Now let's write all of the known information:

$P = 25{,}669.27 \text{ lb}/4 \text{ supports} = 6417.32 \text{ lb per support}$

(*Note:* The load is supported by 2 walls and 2 posts)

$F_c = 1500 \text{ psi}$ (using grade b1 or better from Sample Exercise 8.1 and Table C)

$c = 0.85$ (for round posts)

$K_{ce} = 0.3$ (assuming that the lowest value since finish is not given)

$E = 1{,}800{,}000 \text{ psi}$ (Table C)

$d = 3 \text{ in.}$ (assumed value)

$L_e = 7 \text{ ft } 6 \text{ in.} = 90 \text{ in.}$ (from diagram)

4. Solving for Q we find

$$Q = \frac{(0.3)(1{,}800{,}000)(3 \text{ in.})^2}{(1500)(90)^2}$$

$$= 0.4$$

5. With $P = 6417.32$ lb (required capacity of one post) we solve for A:

$$A = \frac{6417.32 \text{ lb}}{(1500)\left\{\frac{1+0.4}{2(0.85)} - \sqrt{\left[\frac{1+0.4}{2(0.85)}\right]^2 - \frac{0.4}{0.85}}\right\}}$$

$$= \frac{6417.32 \text{ lb}}{(1500)\left(0.82 - \sqrt{(0.82)^2 - 0.47}\right)}$$

$$= 11.56 \text{ in}^2$$

Since the required area for the post cross section is greater than the assumed area of 7.065, we will need to assume a larger size post and recalculate the required area using a large post diameter.

If we repeat the same procedure using 4 in. as the post diameter then the required area of cross section would be computed as 7.23 in², which is smaller than the actual area (12.56 in²) of a 4-in.-diameter post. This would be an appropriate choice for the size of this post.

Problems

1. Compute the moment of inertia about the smallest axis for all standard rectangular wood joist shapes from 2 in. × 4 in. to 2 in. × 12 in.

2. Recompute the moment of inertia for each of the joists in Problem 1 allowing for $\frac{1}{2}$-in. milling along each dimension at least material condition.

3. How much weight can be hung from a 10-ft-long 2-in.-diameter rod made of grade 2 Douglas fir assuming

    (a) standard allowable deflection
    (b) one-half the allowable standard deflection
    (c) twice the allowable standard deflection

4. Determine the actual required size of a Douglas fir select structural grade square beam needed to rest atop three 7-ft 6-in.-tall wood posts located 8 ft apart if the system must carry a load of 55,000 lb.

5. A 12-ft×12-ft steel deck weighing 100 psf is used to support a 40,000 lb load. Compute the size of triangular Douglas fir grade 2 joists needed to support this load if the joists are to be placed

    (a) 24 in. O.C.
    (b) 16 in. O.C.
    (c) 12 in. O.C.

6. The Douglas fir grade 1 joists under an 8-ft-wide boardwalk run lengthwise along the walk 16 in. O.C. Assuming a section load of 20,000 lb, determine the maximum standard length of one section of the walkway if the joists are nominal 2×12's with $\frac{3}{8}$-in. milling at least material condition.

7. Using Douglas fir select structural grade wood determine the size of the posts in Problem 4 if the posts are

   (a) sawn square members
   (b) smooth round members
   (c) square glue-laminated members

8. Draw a detail of the post-and-beam system in Problems 4 and 7. Use notes to indicate the three options listed in Problem 7.

9. Detail the boardwalk section in problem 6 using nominal 2×10 Douglas fir planks for decking run crosswise placed 11 in. O.C. if the planks are milled $\frac{3}{8}$ in. along each dimension at least material condition.

10. Compute the weight of one section of the deck in Problem 9.

11. Compute the total weight and maximum load capacity for the 15-ft×6-ft Douglas fir grade 2 deck shown in Figure 8.16. The deck shown is framed by 2×12 joists milled at $\frac{1}{2}$ in. The deck boards are 2 in.×6 in. run at 8 in. O.C. measured lengthwise along the deck and make a 30° angle with the header joists. The posts are 3 in. in diameter and stand 4 ft high.

FIGURE 8.16

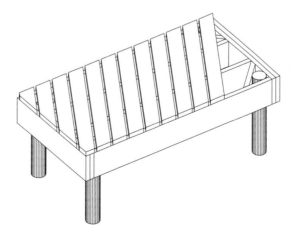

CHAPTER 9

Reinforced Concrete Columns and Footings

9.1 INTRODUCTION

The final chapter of this text focuses on the uses of reinforced concrete as load-bearing members of a structure. Specifically, this chapter explores only the column and footing applications of reinforced concrete. The following topics are addressed as they relate to reinforced concrete:

1. Concrete reinforcing bar
2. Concrete column sizing
   (a) Square
   (b) Round
   (c) Eccentricity
3. Design and detailing for
   (a) Columns
   (b) Footings

9.2 REBAR NOMENCLATURE AND STRENGTH

As with steel and wood, there are certain aspects of reinforced concrete that require a standard nomenclature. The nomenclature for steel concrete reinforcing bar, or **rebar**, as it is commonly called, is discussed in this section. The purpose of rebar is to strengthen a concrete structure from within. The addition of these small steel bars increases the structural integrity of a concrete load-bearing member by increasing its shear capacity and tensile strength.

The method for selecting the grade and size of rebar for a given task is dependent on a number of conditions. We simplify this task by using a standard set of criteria along with a reinforced concrete chart or table. Table D in the back of this text contains data on the standard reinforcing bar sizes. Notice that the pattern of the **ribs** on the outside of each bar is somewhat different, depending on the size of the bar, but the information stamped on each bar is essentially the same. Using Figure 9.1 as a guide let us look at the meaning of each of these pieces of information:

Manufacturer's letters or symbol—Indicates the company that manufactured the bar. This may be a letter or letters, or a symbol. It is always the topmost stamp on a reinforcing bar.

Bar size—A number that is the ASTM designation for the size of the bar. For nominal sizes of various reinforcing bars, refer to Table D.

Steel type—Indicates the type of steel that has been used to make the bar. The options are

S—Billet(A615)

I - Rail(A616)

IR—Supplementary Rail S1(A616)

A—Axle(A617)

W—Low Alloy(A706)

For specific information on the grades and strengths of each type of steel listed above, refer to Table D.

FIGURE 9.1

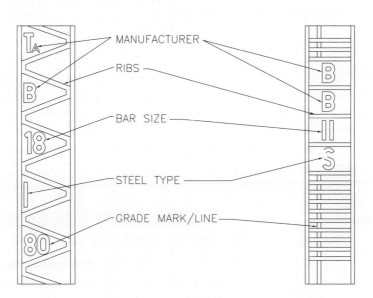

REFER TO TABLE D FOR ADDITIONAL INFORMATION ON REBAR DATA AND SPECIFICATIONS.

Grade Mark—Indicates the grade of steel used. Each type of steel has one or more grades. Higher grades indicate greater yield and tensile strength. When lines are used instead of numbers, follow the grade mark note indicated in Table D to determine the bar grade.

A rebar added to concrete will enhance its structural integrity both laterally (perpendicular to the direction of the rebar) and longitudinally (parallel to the direction of the rebar). In this chapter we are primarily concerned with the lateral effect that rebars have within concrete columns. The strength of a rebar is directly related to its size. The larger the rebar the stronger it is. For the purpose of load computations all grades of steel rebars are assumed to have a modulus of elasticity of 29,000,000 psi.

Note: Look at the bars in Table D. Do you see how the rib patterns differ from bar to bar? This will help you determine the size of the bar. These ribs help to hold the bar in place within the concrete once it has hardened. All bars also have a pair of main ribs on either side. These ribs run lengthwise along the bar and serve to reinforce its ability to resist a bending moment.

Concrete has a relatively high compressive strength, but without steel reinforcement it has a rather low breaking strength. This column has some of its rebars exposed due to external damage. These bars help to strengthen this concrete bridge column laterally. Can you identify the size of these bars by looking at their rib pattern?

9.3 COLUMN ECCENTRICITY

When tall slender objects are compressed along their longest axis they tend to react much like a beam that is loaded along its span. If the load on the end of a column is great enough, then it will cause the column to buckle by making it bow out as shown in Figure 9.2.

Buckling is caused by "weak points" located along the length of a slender object, such as a post or column. No matter how carefully a post or column has been constructed, even the smallest errors or imperfections will lead to the occurrence of weak points. Since a column or post is only as strong as its weakest point it will begin to fail at these points first. To better understand this effect let us look at one of the ways that columns can be loaded.

We will investigate the **concentric** method of loading a column. A column loaded in this manner is subjected to the same load over all 360° of its section perpendicular to the line of action of the force being applied. This principle is illustrated in Figure 9.3. Under ideal conditions a concentrically loaded beam would never fail until it collapsed upon itself under extreme compression.

FIGURE 9.2

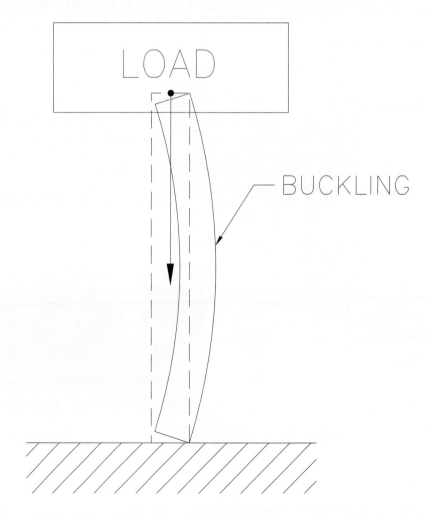

FIGURE 9.3

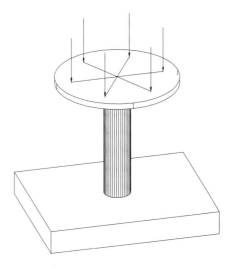

Due to the conditions that cause buckling, no column can be perfectly loaded in a concentric manner. The column will always act as if the load were slightly off center. This effect will create a certain amount of energy that causes the column to bend. The amount of work needed to bend the column in this manner is called its **bending moment**. The bending moment of a column is an indicator of its stability when loaded under a given set of conditions. This principle is illustrated in the left-hand diagram in Figure 9.4, which shows a concentrically loaded column under ideal conditions. A 10,000-lb load placed directly on the centerline of the column causes no moment. The actual effect caused by the 10,000-lb load is shown on the right side of the same figure. The column actually reacts as if the load were placed 3 in. off center. This creates a bending moment of 2500 ft-lb (M = 10,000 lb $\times$ 0.25 ft). The 10,000-lb load is called the **axial compression** or **axial load** because it acts parallel to the axis of the column. Axial compression is designated by the letter P.

FIGURE 9.4

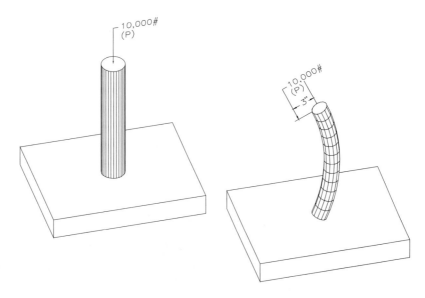

The imaginary distance from the column's center axis at which the axial load is considered to be placed in order to generate the bending moment M is called the **eccentricity** e of the column. If we consider the standard formula for a moment M that is caused by a force F at a given distance d then we recall that

$$M = Fd$$

Replacing F with the value of P for the axial load, and replacing d with e for the distance the load is located from the column axis we find

$$M = Pe$$

Since the value for M in concrete columns is typically measured in kip-ft, P in kips, and e in inches, we need to rewrite this formula to account for the difference in units:

$$M = \frac{Pe}{12}$$

This concrete piling, called a "dead man," is used to elevate the base of the guy wires being used to stabilize this tower because it is located in a flood plain. The "dead man" shown is approximately 10 ft tall and is constantly being subjected to a bending moment. How would you determine the size and reinforcement of this column?

In most instances eccentricity is the unknown variable. For this reason we commonly write the above formula as

$$e = \frac{M \times 12}{P}$$

Now that we have a basic understanding of eccentricity let us discuss how this value is used in the sizing of concrete columns.

9.4 SIZING SQUARE AND ROUND REINFORCED CONCRETE COLUMNS

The sizing of reinforced concrete columns is not a difficult task, but it is one that requires some background information concerning the materials used to construct the column. To size the columns discussed in this chapter we will be using Tables E1 through E4 located at the back of this text. These tables assume the following properties with respect to the columns in question:

Concrete compressive strength of 4 ksi (4000 psi)

Steel rebar yield strength of 60 ksi (60,000 psi) using grade 60 bars

The above conditions are equivalent to those that would exist if the appropriate size grade 60 rebar was used to reinforce a high grade of portland cement or a limestone/sandstone composite. Please be aware that if other grades of concrete and/or rebar are used then the charts in Table E will not be accurate.

Some examples of how reinforced concrete columns are made are shown in Figure 9.5. The left side of the figure shows two examples of square-tied columns, while the right side of the figure shows two examples of round-tied columns. Notice that the rebar is tied together with small amounts of wire.

FIGURE 9.5

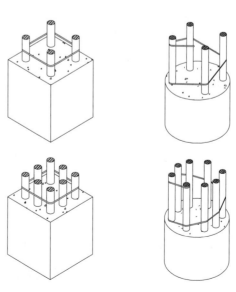

This helps the bars to remain in place while the concrete is being poured, and offers some additional lateral stability to the column.

Looking at the round columns in Figure 9.5 you can see that the bars are tied together a bit differently. The tie wire is run in a spiral, which serves as a more uniform means of holding the bars in place. This is necessary to do in round columns because there are no corners in a circle, therefore, there are no naturally reinforced portions for the shape. Circles also have a smaller cross section than squares circumscribed about the same diameter (see Figure 9.6), which makes the reinforcement provided by the bars and ties even more critical.

The sizing of these columns is relatively easy once you learn to read the charts such as those in Table E. Column sizing is based on three criteria: maximum axial compression P, maximum bending moment M, and column eccentricity e. The actual design load on the column measured in kips is the axial compression P and the generated bending moment measured in kip-ft is the value for M. The sizing charts in Table E are based on the axial load capacity in kips and the eccentricity in inches for the columns shown. The values for M and P are usually known for the column conditions, so it is necessary to find the eccentricity of the column to select the correct size and arrangement.

To illustrate this procedure let us consider a single reinforced concrete column that will be concentrically loaded with a 100,000-lb load while being subjected to a bending moment of 60 kip-ft. Using the formula for eccentricity from Section 9.3 we have

$$e = \frac{60 \times 12}{100}$$
$$= 7.2 \text{ in.}$$

Rounding this value up to the nearest whole number we find $e = 8$ in.

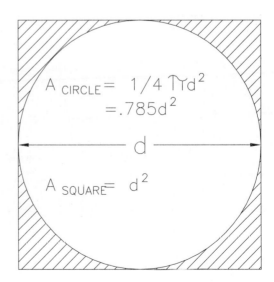

FIGURE 9.6

Now, refering to Table E1 we locate the number 8 along the eccentricity axis (horizontal). Moving up this line to the 100-kip line we find that the two intersect right at the number 11 curve. If we look to the chart in Table E1 we find that the number 11 curve indicates that we need to use a 16-in. square column reinforced by 8 No. 9 rebars. This arrangement would look like the one shown in the lower left portion of Figure 9.5.

If we wanted to use a round column we would need to go to Table E3. Since the 8-in. eccentricity line and the 100 KIP line do not meet in this graph we need to look at the larger graph in Table E4. In Table E4 these lines meet between curves 15 and 16. We would use the uppermost curve (16), which tells us that we need a 20-in.-diameter column with 8 No. 11 bars. This arrangement might look similar to the one shown at the bottom right of Figure 9.5.

Tables E1 and E3 are for smaller square and round columns, respectively, while Tables E2 and E4 are for larger columns. These tables do not contain all of the data for all possible columns, but they do cover a large range of commonly used reinforced concrete columns.

Nonconcentric Loads Considerations

Not all columns will be subjected to concentric loads. Those located at corners or along the exterior walls of a structure will often be subjected to greater bending moments than those columns located throughout the interior of the structure. This is because these columns are not loaded evenly on all sides. In Figure 9.7 the only column that is concentrically loaded is column E. All of the other columns have a resultant direction of bending (indicated by the arrows in the figure) because they are loaded more on one side than on the others. For example, column B is loaded to the east but not to

FIGURE 9.7

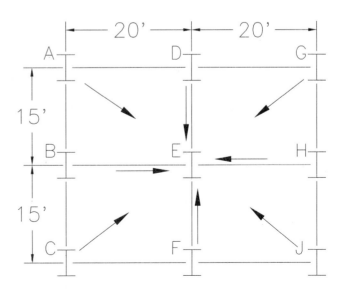

the west. This causes more bending to the east. Column A is loaded to the east and south, but not to the north or west. This causes it to bend more in the southeast direction.

Exterior columns, like exterior beams, tend to carry less of a load than interior columns. This is true because, like beams (discussed in Chapter 6), columns in a rectangular building section carry one-fourth the total load for that section, and columns joining two or more sections carry one-fourth the sum of the loads for all sections they support. The weight distribution for the uniform load of 500 psf on this building section is

$$P_A = 37{,}500 \text{ lb} = 37.5 \text{ kips}$$
$$P_B = 75{,}000 \text{ lb} = 75 \text{ kips}$$
$$P_C = 37{,}500 \text{ lb} = 37.5 \text{ kips}$$
$$P_D = 75{,}000 \text{ lb} = 75 \text{ kips}$$
$$P_E = 150{,}000 \text{ lb} = 150 \text{ kips}$$
$$P_F = 75{,}000 \text{ lb} = 75 \text{ kips}$$
$$P_G = 37{,}500 \text{ lb} = 37.5 \text{ kips}$$
$$P_H = 75{,}000 \text{ lb} = 75 \text{ kips}$$
$$P_J = 37{,}500 \text{ lb} = 37.5 \text{ kips}$$

Maximum bending moments are dependent on a number of conditions that are not discussed in this text. We, therefore, provide these values for you, recognizing that the exterior columns may have a higher value for M than the interior columns. The following bending moments will be used:

$$M_{A,C,G,J} = 60 \text{ kip-ft}$$
$$M_{B,D,F,H} = 50 \text{ kip-ft}$$
$$M_E = 45 \text{ kip-ft}$$

Computing the eccentricity for columns A, C, G, and H we find

$$e = \frac{60 \times 12}{37.5}$$
$$= 19.2 \text{ in.}$$

Rounding up, we'll use 20 in. for eccentricity. These columns carry a load of 37.5 kips each. Using Table E4 we find that curve 20 is the closest choice. This means we'll need to use 24-in.-diameter columns with 8 No. 14 bars for the corners of this structure.

Next, we'll find the eccentricity for columns B, D, F, and H:

$$e = \frac{50 \times 12}{75}$$
$$= 8 \text{ in.}$$

Using Table E4 we find that curve 13 is the appropriate choice. So the remaining exterior columns will be 20 in. in diameter with 4 No. 8 bars.

Finally, we can use the same procedure to size column E:

$$e = \frac{45 \times 12}{150}$$
$$= 3.6 \text{ in.}$$

Rounding up to $e = 4$ in., we find by again using Table E4 that curve 14 represents this column. The center column needs to be 20 in. in diameter with 4 No. 11 bars.

Note that the largest columns in this structure turned out to be the corner columns. This will not always be the case, but this example does show that the sizing of concrete columns can sometimes be unpredictable.

Now let's look at the concrete footings needed to support these columns.

9.5 SIZING FOR COMPRESSIONS ON CONCRETE FOOTINGS

The tremendous load that a concrete column is designed to carry must be distributed as evenly as possible at its base. The device used to do this is called a footing. In this section we discuss two types of footings: the independent square footing and the wall footing.

Figure 9.8 illustrates two views of an independent square footing that could be used to support a square or round column. Independent footings are so named because one column is placed on a single footing. The footing is

FIGURE 9.8

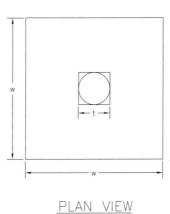

PLAN VIEW

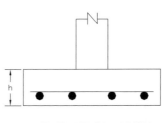

ELEVATION VIEW

sized for the load only from the single column that it supports. The footing in Figure 9.8 is $w \times w$ feet square and h inches high. The dimension t represents the side length of a square column or the diameter of a round column.

To select the correct size independent footing for a particular column you need to know four things:

1. The concrete compressive strength (f_c') in psi
2. The size of the column (t) in inches
3. The soil compression capacity (S_c) in psf
4. The load on the column (k) in kips (same as P in Section 9.4)

Once you have all four of these values then you can use Table F1 to find the appropriate size independent square footing.

Let us consider as our example a square reinforced concrete column that is 10 in. on a side and carries a load of 150,000 lb. If the concrete compressive strength is 2000 psi and the soil pressure capacity is 3000 psf, we need to find a footing for this column. First we consider the concrete strength. This will tell us whether to use the right or left column. Since the concrete strength in this case is 2000 psi, we need to use the left column of Table F1. Second, we note the soil capacity. In this situation we have a soil capacity of 3000 psf so we move down the left column until we reach the section for 3000 psf maximum soil pressure. Third, we find the row that is the next higher value than the one we've been given for allowable load on footing. Since this column carries a load of 150 kips we would go to the row for k = 175 kips. Fourth, we check to see if the column given is large enough to use on this footing. In this case the minimum column width is 10 in., so it is an acceptable column size. Finally, with all of the above criteria satisfied, we note the footing dimensions and rebar information. For this column we would use an 8-ft × 8-ft-square footing 21 in. high with 10 No. 7 bars run each way for reinforcement. The resulting footing is shown in Figure 9.9.

In some structures, especially those that are smaller, independent footings are not required. In such structures the load-bearing elements are the exterior walls themselves and no columns are required. The footing that runs underneath a load-bearing wall is called a **wall footing**. To determine the required dimensions and reinforcement for a wall footing we will use Table F2. The information needed to size a wall footing is almost the same as that for column footings except that the table we'll be using already assumes a concrete strength of 2000 psi. This leaves us with the following required information:

1. The wall thickness t in inches
2. The soil compression capacity S_c in psf
3. The load on the wall footing F in lb/ft
4. The type of wall used (concrete or masonry: For clarification, a concrete wall is a solid poured concrete wall, whereas a masonry wall is a concrete block wall, such as cinder block.)

FIGURE 9.9

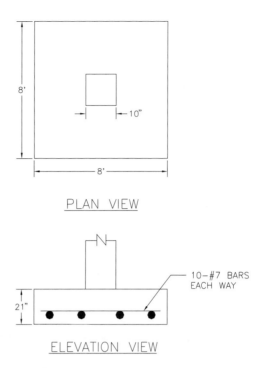

For our example let us look at a 40-ft-long wall that carries a load of 200,000 lb, including the weight of the wall. If the soil capacity is 1500 psf, what size wall and wall footing should be used to support this load?

First we need to find the load in lb/ft on the wall footing. This is found by dividing the total load on the wall by the length of the wall. By doing this we find that the load F is 200,000 lb ÷ 40 ft = 5000 lb/ft. Next we find the section in Table F2 for maximum soil pressure of 1500 psf. The allowable footing loads in this section range from 4125 to 8100 lb/ft. Since the load on this footing is 5000 lb/ft, this grade of soil will be acceptable.

Now we record the necessary information about the required wall and footing. The appropriate selection for this wall footing would be a 10-in.-high by 48-in.-wide footing with 4 No. 4 bars run lengthwise and No. 4 bars run crosswise at 11 in. O.C. The load-bearing wall choices would be a 6-in. or wider concrete wall or a 12-in. or wider masonry wall. The actual allowable load on the wall and footing listed above is 5500 lb/ft. The detail for this wall and wall footing is shown in Figure 9.10.

Keep in mind that many of the values we have assumed in the previous two examples are usually found in charts and tables of soil and concrete. The soil capacity is based on the soil type (such as sand, loam, silt, or clay) and the concrete strength is based on the type and composite used (e.g., portland concrete or cement, limestone, sandstone, or quartz stone conglomerates). These values are available in a number of publications and are often provided for you by a civil or structural engineer.

In the next section we study the methods used to detail the structural elements discussed in this chapter.

FIGURE 9.10

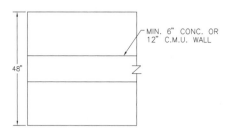

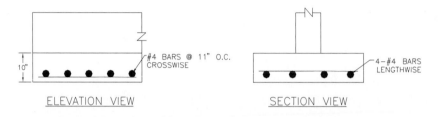

The square reinforced concrete column shown here is used to support a section of a bridge along a major interstate route. Notice that another roadway is located very close to the column. Where do you think the independent column footing is located for this column? What considerations had to be made to size and place the footing?

9.6 DETAILING REINFORCED CONCRETE COLUMNS AND FOOTINGS

Once you have decided on the size and reinforcement of your concrete columns and footings you'll need to transfer that information onto a structural drawing. In this section we offer suggestions and recommend standards to follow for this procedure.

The following specifications should be included in all structural concrete drawings:

Concrete formation—Indicates whether the concrete is field-formed, precast, or formed under specific conditions, such as temperature and setting time.

Concrete strength—The compressive or yield strength of the concrete in psi.

Concrete type—Indicates the material composition of the concrete used to form the structure.

Dimensions—All appropriate dimensions must be shown, including length, width, and height of footings; minimum wall or column size(s) if applicable; rebar location and spacing; and center-to-center column spacing for foundation plans and part plans.

Rebar—Information on rebar grade, size, and type as discussed in Section 9.2 should be included wherever applicable in a nonrepetitious manner.

Special conditions—Any special information that pertains to the preparation, formation, assembly, and/or maintenance of the structure or any of its elements.

Figure 9.11 shows a complete detail of an 8-ft × 8-ft-square footing and the square column that it supports. The two views are necessary for adequate

FIGURE 9.11

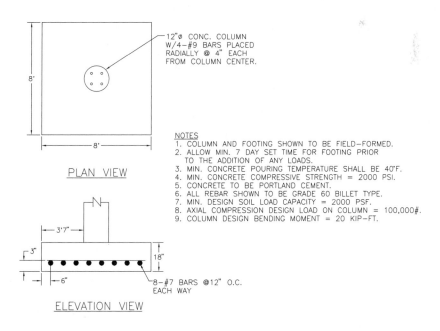

detailing of the reinforcing members. A three-dimensional composite of the same support is shown in Figure 9.12. Pay particular attention to the notes in Figure 9.11. Notice the conditions that are listed for the formation and loading of the column. You may not know these values at the time you design the support, but at least by using this figure as a guide you'll know what items to look for or ask for.

Some helpful hints concerning the location and spacing of rebars are listed below:

1. When rebars are run in both directions in a square footing the bars are usually evenly spaced on center and the distance from the edge of the footing to the outermost bars is usually one-half the O.C. spacing of the bars. Notice this arrangement in the section view of Figure 9.11, where the bars are placed 12-in. O.C. and 6 in. from the edges of the footing. The crossing bars are usually located so that the top sides of the lower bars are four times the bar diameter above the base of the footing. For No. 7 bars, which have a diameter of 1 in., this distance would be 4 in.

2. Rebars located in a poured concrete column are usually placed in a pattern that matches the form being reinforced (square, round, etc.). The bars are usually located about two-thirds of the distance from the center of a round column or at a distance of two-thirds of the side length of a square column apart so that they are equidistant from the column center. Reinforcement for square columns is almost always located in the directions radially from the column center toward the corners first (see Figure 9.5 left side).

3. Ties for rebars are usually small. They are frequently $\frac{1}{8}$ in. to $\frac{1}{4}$ in. in diameter. The gauge values for straight and deformed wire can be found in the *Architectural Graphic Standards*. Square columns generally use individual ties spaced at regular intervals, while round columns use spiral ties with a specified spiral pitch. The computation of tie sizes is not discussed in this text. Wall footings are detailed in a manner similar to that used for independent footings. One major exception is that a top, or plan, view is not generally necessary. The necessary information can usually be indicated on a single section view. The same design and detailing criteria holds true for both wall footings and independent footings. The only additional information needed on a wall footing detail is the size and composition of the load-bearing wall. Also, footing loads, which are indicated in psf for independent footings, need to be indicated in lb/ft for wall footings. Figure 9.13 shows an example of a complete wall footing detail. The three-dimensional composite of this footing is shown in Figure 9.14. The drawings in this section display only some of the many ways that footings may be detailed. Although the actual drawing arrangement may vary from company to company, all of the information discussed in this section should be included or otherwise accounted for in concrete drawings.

FIGURE 9.12

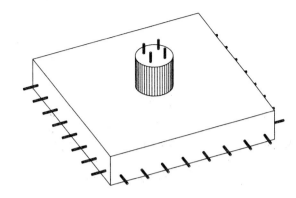

FIGURE 9.13

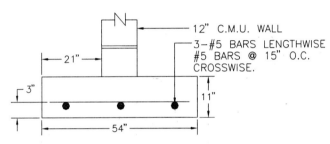

SECTION VIEW

NOTES
1. COLUMN AND FOOTING SHOWN TO BE FIELD-FORMED.
2. ALLOW MIN. 7 DAY SET TIME FOR FOOTING PRIOR TO WALL CONSTRUCTION.
3. MIN. CONCRETE POURING TEMPERATURE SHALL BE 40°F.
4. MIN. CONCRETE COMPRESSIVE STRENGTH = 2000 PSI.
5. CONCRETE TO BE LIMESTONE COMPOSITE.
6. ALL REBAR SHOWN TO BE GRADE 60 BILLET TYPE.
7. MIN. DESIGN SOIL LOAD CAPACITY = 1500 PSF.
8. MAX. DESIGN LOAD ON FOOTING = 6131 LB/FT.

FIGURE 9.14

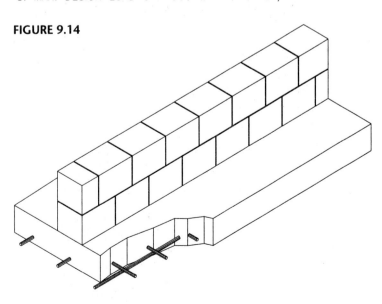

Load-bearing walls are common bridge supports. This wall is being used to support the same bridge as the one shown in the previous photo. A large river is located about 40 ft to the right of this wall. How do you think this fact affected the placement of the wall footing?

9.7 SUMMARY

This chapter has centered around two specific applications for reinforced concrete as structural elements. There are many other uses for reinforced concrete that are not covered in this text, but once you understand the basic principles involved in concrete design you will be able to learn additional applications with greater ease.

Reinforcement bar (rebar) nomenclature has been examined and discussed. Remember that much can be found out about a bar by simply looking at its nomenclature. Also, using rebar charts we can determine such things as size, grade, type, and strength of a bar.

We have also looked at the eccentricity of columns. This helps us to design the column for better stability under varying loads. Remember also that concentrically loaded columns react differently than non-concentrically loaded ones.

The ability to use charts, graphs, and tables for determining design information about concrete columns is essential. In this chapter we have looked at two specific types of reinforced columns for which this method of design is used.

Columns must rest on an adequate support that is designed to evenly distribute its load into the ground. We call this support a footing. Footings may be placed under columns that stand alone or under load-bearing walls. The scope of material covered in this chapter has helped us to understand this concept better. Keep in mind that we have not exhausted all of the possible column and footing arrangements in this text.

Finally, we saw how all of this information is recorded on column and footing details. Again, there are many ways of indicating the necessary information on such a drawing and we have looked at only a few of these. This information and the way it is recorded may differ from company to company, but all of the basic principles that we have discussed still apply.

Now, let us look at some practical applications for the topics discussed in this chapter.

Sample Exercise 9.1

For this example we select an appropriate size column and footing for the center support in Figure 9.15. For the design of this support we assume that all of the weight rests on the center column. Given the dimensions of the building and its weight of 100 lb/ft^2, we determine the total building load to be 240,000 lb or 240 kips. Since we know the bending moment of the column is 100 kip-ft we can determine the eccentricity of this column:

$$e = \frac{100 \times 12}{240}$$
$$= 5 \text{ in.}$$

With $P = 240$ kips and $e = 5$ in we go to Table E2 and find that curve 15 is the correct choice. The specifications for this square column are 20 in. on a side with 8 No. 11 bars.

Now, let's find the specifications for the footing. Since this is an independent square footing, we'll turn to Table F1. Using a 3000-psf soil capacity and 2000-psi concrete strength with a footing load of 240 kips we find that a 10-ft × 10-ft footing 25 in. thick with 11 No. 8 bars run each way is the correct choice.

If you had any trouble understanding the selections made in this problem refer to the procedures listed in Sections 9.4 and 9.5.

FIGURE 9.15

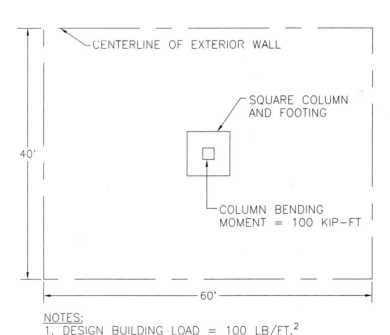

NOTES:
1. DESIGN BUILDING LOAD = 100 LB/FT.^{2}
2. DESIGN CONCRETE STRENGTH = 2000 PSI.
3. DESIGN MIN. SOIL CAPACITY = 3000 PSF.

Sample Exercise 9.2

For this exercise we size the walls and wall footing for the building in Figure 9.15. Let us again consider the total building load of 240,000 lb as the value for the load on the entire wall. Since the perimeter wall is 200 ft at its centerline the load on the wall per foot is 1200 lb/ft. With a maximum soil pressure of 3000 psf we could use the smallest footing in that section. Looking at Table F2, we find that the specifications for this wall and footing are

Dimensions: 10 in. high × 36 in. wide
Bars: 3 No. 4 bars lengthwise and No. 4 bars 10 in. O.C. crosswise
Walls: minimum 6 in. concrete or 12 in. C.M.U.

These specifications may seem appropriate, but further examination shows that this is a grossly oversized support system. Why? The smallest footing and wall in the chart can still carry 2625 lb/ft with soil that is only half as strong as the soil in this example! We see that the smallest footing in this chart would easily handle the load. The criteria for this footing and wall are listed below:

Dimensions: 10 in. high × 36 in. wide
Bars: 3 No. 4 bars lengthwise and No. 3 bars 16 in. O.C. crosswise
Walls: minimum 4 in. concrete or 8 in. C.M.U.

Problems

1. Given the rebar information in the chart below, draw a sketch of each bar. Include all known information and the rib pattern.

	Mfr. symbol	Size	Steel type	Grade
(a)	JL	$1\frac{1}{4}"\ \phi$	Billet	40
(b)	W	$1\frac{7}{8}"\ \phi$	Axle	60
(c)	Mx	$\frac{7}{8}"\ \phi$	Low-alloy	60
(d)	E	$2\frac{1}{2}"\ \phi$	Rail	50
(e)	C	$\frac{7}{16}"\ \phi$	Billet	75

2. Determine the grade of bar needed to carry the maximum tensile load for each of the following conditions:

	Size	Length	Max. elongation
(a)	#11	6 ft	0.064 in.
(b)	#6	6 ft	0.064 in.
(c)	#10	10 ft	0.080 in.
(d)	#7	8 ft	0.250 in.

3. Calculate the eccentricity for each of the columns below:

	Axial compression	Bending moment
(a)	400 kips	200 kip-ft
(b)	25,000 lb	20 kip-ft
(c)	300 tons	150,000 ft-lb
d)	23,500 kg	4000 kg-m

4. Use Tables E1 and E2 to select an appropriate square reinforced concrete column for each column in Problem 3.

5. Use Tables E3 and E4 to select an appropriate round reinforced concrete column for each column in Problem 3.

6. Determine the appropriate dimensions and reinforcement for independent footings to be placed under each column in Problem 4.

7. Draw a detail of each of the footings in Problem 6.

8. A 30-ft × 100-ft warehouse structure weighs 120 lb/ft^2. The structure is designed to carry additional loads of up to 1000 lb/ft^2. Assuming that the concrete used in this structure has a strength of 2000 psi and the soil has a load capacity of 3000 psf, determine the dimensions and reinforcement of the masonry walls and reinforced wall footing for this structure.

9. Draw a detail of the wall and footing found in Problem 8.

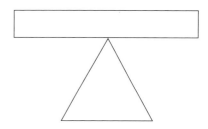

Class Projects

Classroom Project 1

The Strength of Paper

Materials
For this exercise you will be given a paper cup with water and an ordinary sheet of paper.

Purpose
The purpose of this exercise is to investigate the effectiveness of structural elements.

Challenge
Can you think of a way to support the cup of water about 4 in. above the surface of your desk using only the piece of paper you have been given? Try your ideas first (with the cup empty!), then, look at the procedure below to see one way that it can be done.

Procedure

Step I: Hold your piece of paper as shown above in Figure 1A with the longest side vertical.

Step II: Take your paper and fold it in half lengthwise (side to side) as shown in Figure 1B.

Step III: Fold it in half again as shown in Figure 1C, this time folding the paper top to bottom.

Step IV: Fold each side of the paper back toward the center as shown in Figure 1D.

Step V: Fold each end section in half, folding downward as shown in Figure 1E. Your final shape should resemble the one in Figure 1F.

Step VI: Stand your paper on end as shown in Figure 1G.

Step VII: Place the cup (with water in it) carefully on top of the folded paper.

FIGURE 1A

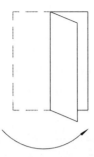

FIGURE 1B

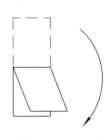

FIGURE 1C

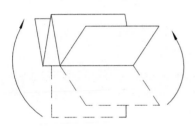

FIGURE 1D

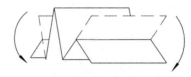

FIGURE 1E

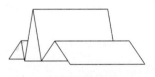

FIGURE 1F

FIGURE 1G

Classroom Project 2

Vector Components

Materials
For this exercise you will be given 2 spring scales and a block of wood with two eye hooks attached to it.

Purpose
The purpose of this exercise is to examine the effects that two forces acting at right angles to one another have on an object.

Challenge
How do you think that two forces acting at right angles to one another will be related to the resulting force that they produce (see Figure 2B)?

Procedure

Step I: You will need to calibrate the spring scale for this exercise. Consult your spring scale instruction manual or your class instructor for this procedure. Then assemble the materials as described below.

Step II: Place the wood block on a flat surface and attach one spring scale as shown in Figure 2A. Now pull the block along the surface for several inches and record the average reading on the scale.

Step III: Next, attach the second spring scale to the other eye hook as shown in Figure 2B. Now pull both spring scales at the same time in the directions indicated in Figure 2B so that the block moves along the "direction of motion" line shown. Record the average reading on each scale.

FIGURE 2A

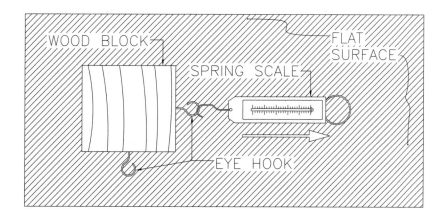

FIGURE 2B

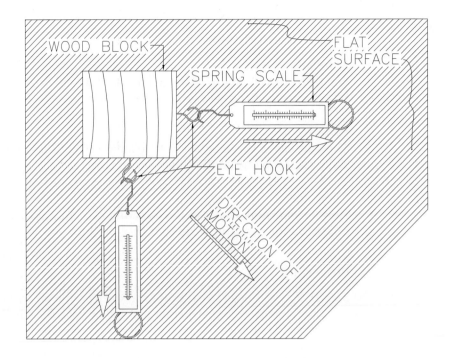

Step IV: Now, test your results by putting each value on a right triangle as shown in Figure 2C. Using this information compute the angle at which your block moved.

Step V: Finally, use the Pythagorean theorem to find the resultant magnitude of the two components from step III. Compare this value to the tension recorded in step II. How should they compare? How do they compare?

FIGURE 2C

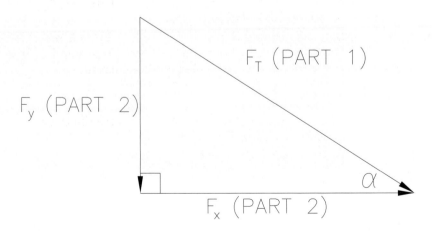

Classroom Project 3

The Breaking Strength of Tissue

Materials
For this exercise you will be given a small, round stone; a spring scale; a metric ruler; and several sheets of tissue paper.

Purpose
Using the items listed above you will determine the approximate breaking strength of the tissue paper in g-cm/cm^2.

Challenge
Can you think of a way that you can measure the approximate breaking strength of a tissue by dropping a rock onto it from a known height?

Procedure

Step I: You will need to calibrate the spring scale for this exercise. Consult your spring scale instruction manual or your class instructor for this procedure. Then assemble the the materials as described below.

Step II: Wrap your stone in one of the pieces of tissue and hang this assembly from the spring scale. Record the weight. Now remove the stone from the tissue and weigh the tissue alone. Subtract the weight of the tissue from the first value you found to determine the weight of the stone.

Step III: Have one person hold a piece of tissue paper firmly several inches above a flat surface. Now, hold the stone at a height of 20 cm above the tissue. Drop the stone onto the tissue.

Step IV: Repeat step III using another tissue folded in half (this doubles the strength of the tissue). Repeat this step until the tissue does not break. *Be sure to hold the stone so that the same side strikes the tissue each time.* Record the number of folded layers of tissue paper at this point.

Step V: Determine the approximate surface area of the side of your stone that struck the tissue. Record this value.

Step VI: Calculate the approximate kinetic energy at impact with the tissue using $KE = Fd$.

Step VII: Divide your result in step VI by your result in step V. Now divide that result by the number of folds from step IV. This is the breaking strength of your tissue paper in g-cm/cm^2.

Can you think of a procedure to measure the tissue's breaking strength more accurately? Describe it.

Classroom Project 4

Angled Acting Forces

Materials
For this project you will be given a paper clip, a small protractor, a rubber band, a spring scale, a pen of some type, and a small object.

Purpose
The purpose of this exercise is to investigate the various ways in which angled acting forces need to be applied to counteract a moment caused by gravity.

Challenge
How do you think that adjusting the angle of an acting force on a moment arm will change the resulting moment?

Procedure

Step I: You will need to calibrate the spring scale for this exercise. Consult your spring scale instruction manual or your class instructor for this procedure. Then assemble the materials as shown in Figure 4A. *Do not attach the weight and spring scale until step III.*

FIGURE 4A

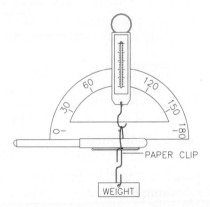

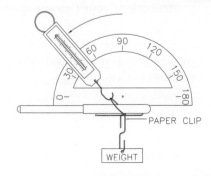

Step II: Determine the weight of the object given to you using the spring scale. Be sure to record this value.

Step III: Now, attach the weight and spring scale to the assembly as shown in the figure. Holding the pen *loosely* with one hand, hold the spring scale vertically with the other hand so that the weight hangs freely with the pen parallel to the ground (see Figure 4A, top).

Step IV: With the weight hanging freely and the pen resting horizontally, begin pulling the top of the spring scale toward your right hand until you are pulling at roughly a 45° angle (see Figure 4A, bottom). Record the reading on the spring scale (be sure that the pen remains level during your reading).

Step V: Continue to pull the spring scale back until it is at approximately a 30° angle. *Remember to grip the pen loosely; gripping the pen too tightly will result in a false reading on the spring scale.* Record the new reading on the spring scale. Do you see a pattern developing?

Step VI: Now pull the spring scale as close to horizontal as you can without exceeding the highest reading on the scale. Record this value and the angle at which you are pulling.

Step VII: Finally, compare each result that you recorded with the value you obtain by using the formula $F_T = F_N \div \sin \psi$, where F_T = force along the line of force (spring scale reading), F_N = normal force against the line of interest (hanging weight), ψ = angle that line of force (F_T) makes with line of interest (pen). Do all of the results match the formula values. If not, why not?

Classroom Project 5

Linear Loads Distribution

Materials
For this exercise you will be given a ruler, 2 spring scales, 3 large clips, a paper clip, and a weighted object.

Purpose
You will use the tools listed above to investigate the effects caused by locating an independent external load on a beam supported at either end.

Challenge
How do you think that the relocation of external loads on a beam will affect the forces on its supporting members?

Procedure

Step I: You will need to calibrate the spring scale for this exercise. Consult your spring scale instruction manual or your class instructor for this procedure. Then assemble the the materials as described below.

FIGURE 5A

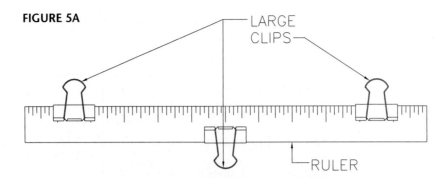

Step II: Assemble the ruler and clips as shown in Figure 5A. Make sure that the top clips are equidistant from the center of the ruler. Be sure, also, to center the bottom clip on the ruler.

Step III: Using one spring scale and a paper clip, weigh the weighted object. Record this value. Now, attach the object to the bottom clip using the paper clip. Then attach both spring scales to the top clips. Your assembly should resemble the one shown in Figure 5B.

Hold the two spring scales up so that the entire ruler/clip/object assembly hangs freely and the ruler is parallel to the ground. Record the values shown on each of the spring scales (they should be equal).

FIGURE 5B

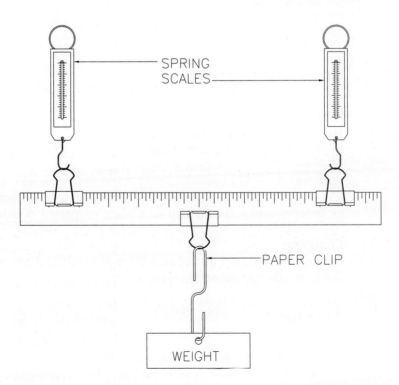

Step IV: Add the values from the two spring scales in step III. Subtract the weight of the object from this value. The difference is the weight of the ruler and clips. Record this value.

Step V: Now, move the weighed object and bottom clip so that they are one-third of the way along the ruler between the spring scales. Hold the spring scales up as you did in step III and record the value you now read (the values will not be equal). Repeat this procedure three more times using different locations along the ruler to hang the weight. Be sure to record the distance that the weight is from each spring scale and the reading on each spring scale each time.

Step VI: In each of the previous cases, notice the distance that the clip/weight assembly was from the left spring scale. Divide this distance by the total distance between both spring scales. Now multiply this result by the weight of the object you found in the early portion of step III. Compare your answer to the *actual* reading on the right spring scale.

Now, subtract half the weight of the ruler/clip assembly you found in step IV from the value you read on the right spring scale. How does this value compare to the *computed* value you found above?

Classroom Project 6

Looking at Flexure

Materials
For this exercise you will be given 2 popsicle sticks, a paper towel, and a cup of water.

Purpose
Using the materials listed above you will investigate how flexure occurs in a beam under stress.

Challenge
What do you think happens to a beam at the point where it bends under a load? What happens to the top and bottom surfaces where it bends?

Procedure

Step I: Soak one popsicle stick in the cup of water for at least 5 minutes. Meanwhile, take the dry popsicle stick and, grasping it near each end, bend it slowly until it begins to splinter (it may snap suddenly—that's OK). Look at the place where it began to break. Record what you see.

Step II: Now, take the wet popsicle stick and wipe the excess water off with the paper towel. Repeat the procedure in step I using the wet stick, making sure to bend the stick slowly (it will be more flexible that the dry one). Bend the stick only until it begins to splinter. What do you see?

Now, bend the stick carefully some more. Do you see what is happening on both sides of the stick? Record your observations.

Step III: You have just witnessed flexure. Write a definition for this term based on what you have just observed.

Classroom Project 7

Hooke's Law in Action

Materials
For this exercise you will be given a pencil, a rubber band, a spring scale, and a piece of paper.

Purpose
Using the materials listed above you will determine how uniformly increasing loads affect the elongation of the rubber band.

Challenge
How do you think your rubber band will stretch under increasing loads? Will there be any "sticking points," or will the rubber band stretch faster at different levels of stress?

Procedure

Step I: You will need to calibrate the spring scale for this exercise. Consult your spring scale instruction manual or your class instructor for this procedure. Then continue with the procedure in the steps that follow.

Step II: Now place the rubber band onto the hook at the end of the spring scale, and lay the scale face up on top of the paper so that you can read the numbers (see Figure 7A, top).

Step III: Insert the pencil into the rubber band as shown in the diagram and pull it to the point at which the rubber band just begins to stretch (spring scale should read "zero" at this point). Make a mark with the pencil at this point.

Step IV: Begin stretching the rubber band while marking at regular intervals the spot (see Figure 7A, bottom) to which it stretches (example: 50 g, 100 g, 150 g).

Step V: After you reach the highest point of tension on the scale take your ruler and measure the distance between each pencil mark.

Hooke's law states that uniform deformation occurs under uniformly increasing stresses. How does this law apply to the project you just completed? Did your results reflect this statement? Why or why not?

FIGURE 7A

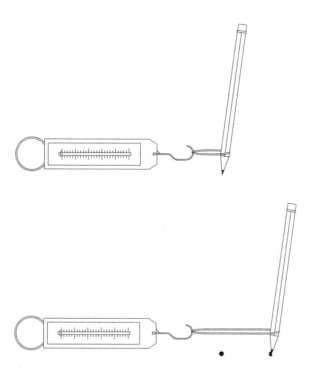

Classroom Project 8

The Strength of Shapes

Materials

For this exercise you will be given 6 popsicle sticks, a heavy string, a dowel stick or metal rod, a ruler, several small weights, and two blocks with grooves in them like the one shown in Figure 8A.

Purpose

Using these items you will investigate how the shape of a beam affects its strength.

Challenge

Have you ever wondered why beams and columns come in so many different shapes? Houses, buildings, bridges, and towers have many different appearances and are often made up of a variety of different shapes. Do you know what shapes are the strongest?

Procedure

Step 1: Figure 8A is an illustration of one of the two wood blocks that you will receive. Each groove has been labeled with a number. Throughout this exercise you will be inserting popsicle sticks into these grooves and hanging weights from them.

FIGURE 8A

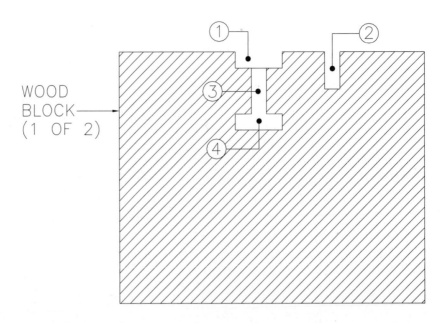

To prepare for this exercise place the two wood blocks about $3\frac{1}{2}$ in. apart with the groove openings facing each other. Now prepare the string and dowel by tying the string to the rod/dowel using a cat's paw as shown in Figure 8B. Now tie the two loose ends together and loop the far end over one of the popsicle sticks (Figure 8C). Try to keep the knot off to one side of the stick as illustrated in Figure 8D.

Step II: Now, place the popsicle stick on the blocks in groove 1 (see Figure 8A). Slip the rod/dowel through the hole in one of the weights as shown in Figure 8E. Slowly lower the weight until it hangs freely from the popsicle stick. Continue adding weights until the popsicle stick splinters or breaks. Record the amount of weight on the string at this point.

Step III: Repeat step II, using two popsicle sticks together placed in groove 1. Then repeat step II, placing one popsicle stick in groove 2. Finally, repeat step II with popsicle sticks in grooves 1, 3, and 4 at the same time (I-shape). In each case record your findings.

Step IV: Look at the depth and area of the cross section of each beam shape you just tested. Compare these values with the amount of weight that each shape held. What do you notice?

FIGURE 8B

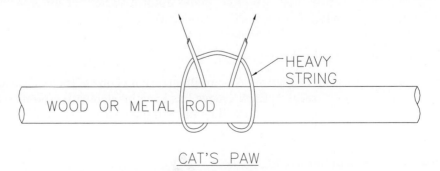

CLASS PROJECTS △ **243**

FIGURE 8C

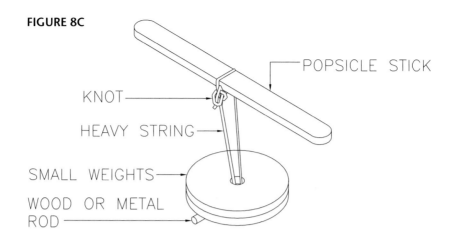

FIGURE 8D

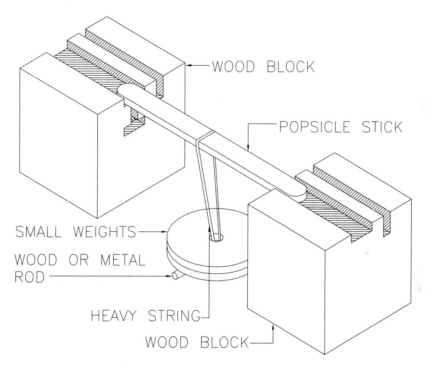

FIGURE 8E

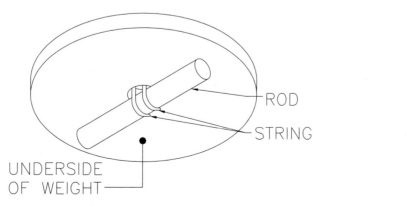

Classroom Project 9

Column Eccentricity

Materials
For this exercise you will be given a dowel stick with a flat piece of wood attached to one end, a thick rubber band, a spring scale, and a metric ruler.

Purpose
Using these tools you will investigate the factors that contribute to the bending moment of a column.

Challenge
Do you know why poles, posts, and columns are likely to bend when forces compress their ends together?

Procedure

Step I: You will need to calibrate the spring scale for this exercise. Consult your spring scale instruction manual or your class instructor for this procedure. Then assemble the the materials as described below.

Step II: Hold the dowel vertically against a flat surface at its base (as indicated in Figure 9A). Loop the rubber band 10 cm from the center of the dowel on the flat piece of wood as shown in Figure 9A.

Step III: Now hook the spring scale to the rubber band as shown in Figure 9A (you may double the rubber band if it stretches too much). Pull downward on the spring scale until it reads 50 g. Record the approximate number of millimeters the dowel bent. Repeat this procedure at tensions of 100 and 150 g. Record the approximate amount of bend in each case.

FIGURE 9A

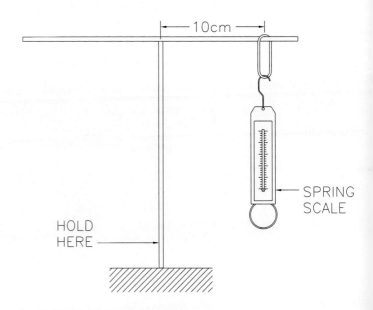

FIGURE 9B

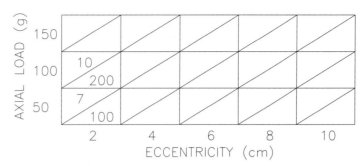

NOTE: VALUES IN BOXES INDICATE AMOUNT OF BEND IN mm OVER BENDING MOMENT IN g-cm.

Construct a chart that looks like the one in Figure 9B. Record your results for millimeters of bend in the upper left half of each box in the 10-cm column since your spring scale is pulling from a distance of 10-cm from the dowel.

Step IV: Compute the bending moment for each force and location listed in your chart by multiplying the tension on the spring scale in grams by the distance from the dowel in centimeters. Your results in each case will be the bending moment in g-cm (gram-centimeters). Record each of these values in the lower right half of each box of your chart.

Step V: Repeat step III four more times, each time moving the rubber band 2 cm closer to the dowel. Continue to record your findings in your chart. When you have finished, the chart will be full. Examine the values in your completed chart. Compare the amount of bend in each case to the value of the bending moment. What do you notice?

Step V: Now hold the dowel vertically as you did in step II, but remove the spring scale and rubber band. Place your other hand directly on top of the dowel and press downward. What happens to the dowel as you increase the pressure? Why?

Term Project 1

Materials
    12 popsicle sticks
    2 standard size rubber bands (not wide)
    1 pen spring
    1 bottle of Elmer's Glue-all ($1\frac{1}{4}$ oz)
    3 feet of string or sewing thread

Objective
You will be required to construct a catapult that will attach to a sawhorse using only the materials listed above so that it can launch a golf ball a maximum distance.

Criteria
1. The catapult arm may not exceed 12 in. in length.
2. The catapult arm must rest parallel to the ground when fully loaded and armed.
3. The fulcrum need not be located at the end of the catapult arm.
4. The catapult need not remain attached to the sawhorse after launch.
5. Glue lamination of popsicle sticks is not allowed. Use only the amount of glue necessary to secure the sticks to each other.
6. The base of the structure must continue to touch the sawhorse until the golf ball has been released.
7. The launch must be activated remotely by the string. No one may be touching the catapult arm, fulcrum, or sawhorse during the launch.
8. The string release mechanism may not provide any additional mechanical advantage to the catapult mechanism.

Grading
Grading will be based on the following criteria:
1. Conformity to criteria (all projects must conform to the minimum criteria to be considered for eligibility)
2. Originality
3. Energy output of the catapult
4. Application of static and kinetic physics principles

Term Project 2

Materials:
100 sheets of notebook paper $8\frac{1}{2}$ in. $\times$ 11 in.
One box of 250 round toothpicks
1 bottle of Elmer's Glue-all ($1\frac{1}{4}$ oz)

Objective
You will be required to construct a round or rectangular pedestal table using only the materials listed above so that it can hold a maximum amount of weight.

Criteria
1. The top surface of the table must be a true circle or square.
2. The load-bearing surface top must rest parallel to the ground.
3. The pedestal support must be at least 10 in. high.
4. Only one pedestal support may be used.
5. Glue lamination of toothpicks or paper is not allowed. Use only the amount of glue necessary to secure the toothpicks and paper to each other.

6. The base of the structure must continue to touch the ground and remain level while stressed.
7. The base of the table must be a true circle or square of monolithic type that is removable from the table to allow inspection of the interior of the pedestal support structure.
8. A project will be considered to be at yield (structural failure) when
   (a) it has collapsed.
   (b) it receives irreparable structural damage.
   (c) it is no longer touching the ground in accordance with item 6 of this section.

Grading
Grading will be based on the following criteria:
1. Conformity to criteria (all projects must conform to the minimum criteria to be considered for eligibility)
2. Originality
3. Strength of structure
4. Application of static physics principles

Term Project 3

Materials:
   One box of 250 round toothpicks
   1 bottle of Elmer's Glue-all ($1\frac{1}{4}$ oz)

Objective
You will be required to construct a rectangular platform using only the materials listed above so that it can hold a maximum amount of weight.

Criteria
1. The rectangular platform must be a true rectangle having a length that is at least 1.5 times the width.
2. The load-bearing platform top must rest parallel to the ground.
3. The platform supports must be at least one-half the length of the longest side of the rectangular platform.
4. Four supports must be used and they must be placed at least three-fourths of the length of the shortest side of the rectangular platform apart from one another.
5. Glue lamination of toothpicks is not allowed. Use only the amount of glue necessary to secure the toothpicks to each other.
6. The structure must continue to touch the ground at four and only four points while stressed.

7. A project will be considered to be at yield (structural failure) when
   (a) it has collapsed.
   (b) it receives irreparable structural damage.
   (c) it is no longer touching the ground in accordance with item 6 of this section.

Grading
1. Grading will be based on the following criteria:
   (a) Conformity to criteria (all projects must conform to the minimum criteria to be considered for eligibility)
   (b) Originality
   (c) Strength of structure
   (d) Application of static physics principles

Term Project 4

Materials
One box of 250 round toothpicks
10 ft of fine sewing thread
1 bottle of Elmer's Glue-all ($1\frac{1}{4}$ oz)

Objective
You will be required to construct a bridge using only the materials listed above so that it can hold a maximum amount of weight suspended from its center.

Criteria
1. The bridge must be a minimum of 10 in. in length (the bridge will rest on two blocks spaced 10 in. apart).
2. The load-bearing surface of the bridge must rest parallel to the ground.
3. Any structure(s) located beneath the bridge may not extend more than 3 in. below the bridge load-bearing surface.
4. The load bearing surface may not be more than 4 toothpicks in thickness.
5. Glue lamination of toothpicks is not allowed. Use only the amount of glue necessary to secure the toothpicks to each other.
6. The structure must continue to touch the blocks at all support points when stressed.
7. A project will be considered to be at yield (structural failure) when
   (a) it has collapsed.
   (b) it receives irreparable structural damage.
   (c) it is no longer touching the blocks in accordance with item 6 of this section.

Grading
1. Grading will be based on the following criteria:
   (a) Conformity to criteria (all projects must conform to the minimum criteria to be considered for eligibility)
   (b) Originality
   (c) Strength of structure
   (d) Application of static physics principles

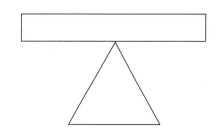

Tables

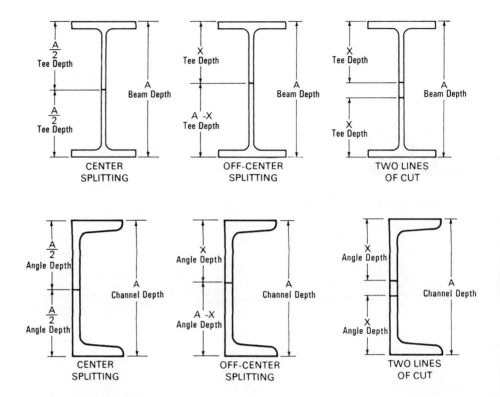

Structural Tees and Angles
Split from Beams and Channels

Note: Tees will be split by burning or by rotary saw at mill option.

Depth Tolerances

Nominal Depth of Beam or Channel, in.	Split Depth Tolerance Over or Under, in.
*4, To 6, excl	1/8
6 to 16, excl	3/16
16 to 20, excl	1/4
20 to 24, excl	5/16
24 and over	3/8

Tolerances for depth of tees or angles include allowable tolerances in depth for the beams or channels before splitting.

*Tolerances for 3 in. sections are subject to inquiry.

Other Tolerances

Dimensions and Length: Tolerances for the beams or channels from which the tees or angles are split will apply.

Straightness:
$$\text{Straightness} = 1/8 \text{ in.} \times \frac{\text{length in ft}}{5 \text{ ft}}$$

Taken from Bethlehem Steel *Structural Shapes* handbook.

SYMBOLS

- **A** Cross sectional area (sq in.)
- **E** Modulus of elasticity of steel (29,000 ksi)
- **E_o** Distance from the center of the web to the shear center of a channel section (in.)
- **F_y** Specified minimum yield stress (ksi)
- **F'_y** Theoretical yield stress at which the shape becomes noncompact as defined by flange criteria (ksi)
- **I_x, I_y** Moment of inertia of a section (in.^{4})
- **M_P** Plastic moment (ft-kips)
- **P_y** Plastic axial load: equal to the profile area times specified minimum yield stress (kips)
- **R** Radius of fillet (in.)
- **S_x, S_y** Elastic section modulus (in.^{3}) (based on the exact theoretical value of I)
- **T** Tangent distance on the web between fillets (in.)
- **Z_x,** Plastic section modulus (in.^{3})
- **a** Distance from web face to edge of flange (in.)
- **b_f** Width of flange (in.)
- **d** Depth of section (in.)
- **g** Usual gage in flange (in.)
- **k** Distance from outside of flange face to intersection of fillet with web (in.)
- **k_1** Distance from center line of web to intersection of fillet with flange (in.)
- **r_x, r_y** Radius of gyration (in.)
- **r_T** Radius of gyration about the Y-Y axis of a "T-section" consisting of the compression flange and fillets plus one-sixth of the web's clear distance between the flanges (in.). For beams with sloping flanges, the web's clear depth is the distance between the points at which the planes of the insides of the flanges intersect the plane of the face of the web.
- **t_f** Flange thickness for shapes with no flange slope; also, average thickness for shapes with sloped flanges (in.)
- **t_w** Web thickness (in.)
- **x,y** Distances from outside face of section to neutral axes Y-Y and X-X respectively (in.)

Taken from Bethlehem Steel *Structural Shapes* handbook.

WIDE FLANGE SHAPES

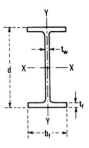

Theoretical Dimensions and Properties for **Designing**

Section Number	Weight per Foot	Area of Section	Depth of Section	Flange		Web Thickness	Axis X-X			Axis Y-Y			r_T
		A	d	Width b_f	Thickness t_f	t_w	I_x	S_x	r_x	I_y	S_y	r_y	
	lb	in.²	in.	in.	in.	in.	in.⁴	in.³	in.	in.⁴	in.³	in.	in.
W36 x 300	300	88.3	36.74	16.655	1.680	0.945	20300	1110	15.2	1300	156	3.83	4.39
280	280	82.4	36.52	16.595	1.570	0.885	18900	1030	15.1	1200	144	3.81	4.37
260	260	76.5	36.26	16.550	1.440	0.840	17300	953	15.0	1090	132	3.78	4.34
245	245	72.1	36.08	16.510	1.350	0.800	16100	895	15.0	1010	123	3.75	4.32
230	230	67.6	35.90	16.470	1.260	0.760	15000	837	14.9	940	114	3.73	4.30
W36 x 210	210	61.8	36.69	12.180	1.360	0.830	13200	719	14.6	411	67.5	2.58	3.09
194	194	57.0	36.49	12.115	1.260	0.765	12100	664	14.6	375	61.9	2.56	3.07
182	182	53.6	36.33	12.075	1.180	0.725	11300	623	14.5	347	57.6	2.55	3.05
170	170	50.0	36.17	12.030	1.100	0.680	10500	580	14.5	320	53.2	2.53	3.04
160	160	47.0	36.01	12.000	1.020	0.650	9750	542	14.4	295	49.1	2.50	3.02
150	150	44.2	35.85	11.975	0.940	0.625	9040	504	14.3	270	45.1	2.47	2.99
135	135	39.7	35.55	11.950	0.790	0.600	7800	439	14.0	225	37.7	2.38	2.93
W33 x 241	241	70.9	34.18	15.860	1.400	0.830	14200	829	14.1	932	118	3.63	4.17
221	221	65.0	33.93	15.805	1.275	0.775	12800	757	14.1	840	106	3.59	4.15
201	201	59.1	33.68	15.745	1.150	0.715	11500	684	14.0	749	95.2	3.56	4.12
W33 x 152	152	44.7	33.49	11.565	1.055	0.635	8160	487	13.5	273	47.2	2.47	2.94
141	141	41.6	33.30	11.535	0.960	0.605	7450	448	13.4	246	42.7	2.43	2.92
130	130	38.3	33.09	11.510	0.855	0.580	6710	406	13.2	218	37.9	2.39	2.88
118	118	34.7	32.86	11.480	0.740	0.550	5900	359	13.0	187	32.6	2.32	2.84
W30 x 211	211	62.0	30.94	15.105	1.315	0.775	10300	663	12.9	757	100	3.49	3.99
191	191	56.1	30.68	15.040	1.185	0.710	9170	598	12.8	673	89.5	3.46	3.97
173	173	50.8	30.44	14.985	1.065	0.655	8200	539	12.7	598	79.8	3.43	3.94
W30 x 132	132	38.9	30.31	10.545	1.000	0.615	5770	380	12.2	196	37.2	2.25	2.68
124	124	36.5	30.17	10.515	0.930	0.585	5360	355	12.1	181	34.4	2.23	2.66
116	116	34.2	30.01	10.495	0.850	0.565	4930	329	12.0	164	31.3	2.19	2.64
108	108	31.7	29.83	10.475	0.760	0.545	4470	299	11.9	146	27.9	2.15	2.61
99	99	29.1	29.65	10.450	0.670	0.520	3990	269	11.7	128	24.5	2.10	2.57

All shapes on these pages have parallel-faced flanges.

Taken from Bethlehem Steel *Structural Shapes* handbook.

WIDE FLANGE SHAPES

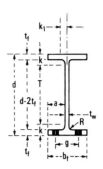

Approximate Dimensions for **Detailing**

Section Number	Weight per Foot	Depth of Section	Flange		Web Thickness	Half Web Thickness	d-2t$_f$	a	T	k	k$_1$	R	Usual Flange Gage
			Width	Thickness									
		d	b$_f$	t$_f$	t$_w$	$\frac{t_w}{2}$							g
	lb	in.	in.	in.	in.	in.	in.	in.	in.	in.	in.	in.	in.
W36 x	300	36¾	16⅝	1 11/16	15/16	½	33⅜	7⅞	31⅛	2 13/16	1½	0.95	5½
	280	36½	16⅝	1 9/16	⅞	7/16	33⅜	7⅞	31⅛	2 11/16	1½	0.95	5½
	260	36¼	16½	1 7/16	13/16	7/16	33⅜	7⅞	31⅛	2 9/16	1½	0.95	5½
	245	36⅛	16½	1⅜	13/16	7/16	33⅜	7⅞	31⅛	2½	1 7/16	0.95	5½
	230	35⅞	16½	1¼	¾	⅜	33⅜	7⅞	31⅛	2⅜	1 7/16	0.95	5½
W36 x	210	36¾	12⅛	1⅜	13/16	7/16	34	5⅝	32⅛	2 5/16	1¼	0.75	5½
	194	36½	12⅛	1¼	¾	⅜	34	5⅝	32⅛	2 3/16	1 3/16	0.75	5½
	182	36⅜	12⅛	1 3/16	¾	⅜	34	5⅝	32⅛	2⅛	1 3/16	0.75	5½
	170	36⅛	12	1⅛	11/16	⅜	34	5⅝	32⅛	2	1 3/16	0.75	5½
	160	36	12	1	⅝	5/16	34	5⅝	32⅛	1 15/16	1⅛	0.75	5½
	150	35⅞	12	15/16	⅝	5/16	34	5⅝	32⅛	1⅞	1⅛	0.75	5½
	135	35½	12	13/16	⅝	5/16	34	5⅝	32⅛	1 11/16	1⅛	0.75	5½
W33 x	241	34⅛	15⅞	1⅜	13/16	7/16	31⅜	7½	29¾	2 3/16	1 3/16	0.70	5½
	221	33⅞	15¾	1¼	¾	⅜	31⅜	7½	29¾	2 1/16	1 3/16	0.70	5½
	201	33⅝	15¾	1⅛	11/16	⅜	31⅜	7½	29¾	1 15/16	1⅛	0.70	5½
W33 x	152	33½	11⅝	1 1/16	⅝	5/16	31⅜	5½	29¾	1⅞	1⅛	0.70	5½
	141	33¼	11½	15/16	⅝	5/16	31⅜	5½	29¾	1¾	1 1/16	0.70	5½
	130	33⅛	11½	⅞	9/16	5/16	31⅜	5½	29¾	1 11/16	1 1/16	0.70	5½
	118	32⅞	11½	¾	9/16	5/16	31⅜	5½	29¾	1 9/16	1 1/16	0.70	5½
W30 x	211	31	15⅛	1 5/16	¾	⅜	28 5/16	7⅛	26¾	2⅛	1⅛	0.65	5½
	191	30⅝	15	1 3/16	11/16	⅜	28 5/16	7⅛	26¾	1 15/16	1 1/16	0.65	5½
	173	30½	15	1 1/16	⅝	5/16	28 5/16	7⅛	26¾	1⅞	1 1/16	0.65	5½
W30 x	132	30¼	10½	1	⅝	5/16	28 5/16	5	26¾	1¾	1 1/16	0.65	5½
	124	30⅛	10½	15/16	9/16	5/16	28 5/16	5	26¾	1 11/16	1	0.65	5½
	116	30	10½	⅞	9/16	5/16	28 5/16	5	26¾	1⅝	1	0.65	5½
	108	29⅞	10½	¾	9/16	5/16	28 5/16	5	26¾	1 9/16	1	0.65	5½
	99	29⅝	10½	11/16	½	¼	28 5/16	5	26¾	1 7/16	1	0.65	5½

Taken from Bethlehem Steel *Structural Shapes* handbook.

WIDE FLANGE SHAPES

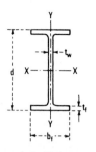

Theoretical Dimensions and Properties for **Designing**

Section Number	Weight per Foot	Area of Section A	Depth of Section d	Flange Width b_f	Flange Thickness t_f	Web Thickness t_w	Axis X-X I_x	Axis X-X S_x	Axis X-X r_x	Axis Y-Y I_y	Axis Y-Y S_y	Axis Y-Y r_y	r_T
	lb	in.²	in.	in.	in.	in.	in.⁴	in.³	in.	in.⁴	in.³	in.	in.
W27 x	178	52.3	27.81	14.085	1.190	0.725	6990	502	11.6	555	78.8	3.26	3.72
	161	47.4	27.59	14.020	1.080	0.660	6280	455	11.5	497	70.9	3.24	3.70
	146	42.9	27.38	13.965	0.975	0.605	5630	411	11.4	443	63.5	3.21	3.68
W27 x	114	33.5	27.29	10.070	0.930	0.570	4090	299	11.0	159	31.5	2.18	2.58
	102	30.0	27.09	10.015	0.830	0.515	3620	267	11.0	139	27.8	2.15	2.56
	94	27.7	26.92	9.990	0.745	0.490	3270	243	10.9	124	24.8	2.12	2.53
	84	24.8	26.71	9.960	0.640	0.460	2850	213	10.7	106	21.2	2.07	2.49
W24 x	162	47.7	25.00	12.955	1.220	0.705	5170	414	10.4	443	68.4	3.05	3.45
	146	43.0	24.74	12.900	1.090	0.650	4580	371	10.3	391	60.5	3.01	3.43
	131	38.5	24.48	12.855	0.960	0.605	4020	329	10.2	340	53.0	2.97	3.40
	117	34.4	24.26	12.800	0.850	0.550	3540	291	10.1	297	46.5	2.94	3.37
	104	30.6	24.06	12.750	0.750	0.500	3100	258	10.1	259	40.7	2.91	3.35
W24 x	94	27.7	24.31	9.065	0.875	0.515	2700	222	9.87	109	24.0	1.98	2.33
	84	24.7	24.10	9.020	0.770	0.470	2370	196	9.79	94.4	20.9	1.95	2.31
	76	22.4	23.92	8.990	0.680	0.440	2100	176	9.69	82.5	18.4	1.92	2.29
	68	20.1	23.73	8.965	0.585	0.415	1830	154	9.55	70.4	15.7	1.87	2.26
W24 x	62	18.2	23.74	7.040	0.590	0.430	1550	131	9.23	34.5	9.80	1.38	1.71
	55	16.2	23.57	7.005	0.505	0.395	1350	114	9.11	29.1	8.30	1.34	1.68
W21 x	147	43.2	22.06	12.510	1.150	0.720	3630	329	9.17	376	60.1	2.95	3.34
	132	38.8	21.83	12.440	1.035	0.650	3220	295	9.12	333	53.5	2.93	3.31
	122	35.9	21.68	12.390	0.960	0.600	2960	273	9.09	305	49.2	2.92	3.30
	111	32.7	21.51	12.340	0.875	0.550	2670	249	9.05	274	44.5	2.90	3.28
	101	29.8	21.36	12.290	0.800	0.500	2420	227	9.02	248	40.3	2.89	3.27
W21 x	93	27.3	21.62	8.420	0.930	0.580	2070	192	8.70	92.9	22.1	1.84	2.17
	83	24.3	21.43	8.355	0.835	0.515	1830	171	8.67	81.4	19.5	1.83	2.15
	73	21.5	21.24	8.295	0.740	0.455	1600	151	8.64	70.6	17.0	1.81	2.13
	68	20.0	21.13	8.270	0.685	0.430	1480	140	8.60	64.7	15.7	1.80	2.12
	62	18.3	20.99	8.240	0.615	0.400	1330	127	8.54	57.5	13.9	1.77	2.10
W21 x	57	16.7	21.06	6.555	0.650	0.405	1170	111	8.36	30.6	9.35	1.35	1.64
	50	14.7	20.83	6.530	0.535	0.380	984	94.5	8.18	24.9	7.64	1.30	1.60
	44	13.0	20.66	6.500	0.450	0.350	843	81.6	8.06	20.7	6.36	1.26	1.57

All shapes on these pages have parallel-faced flanges.

Taken from Bethlehem Steel *Structural Shapes* handbook.

WIDE FLANGE SHAPES

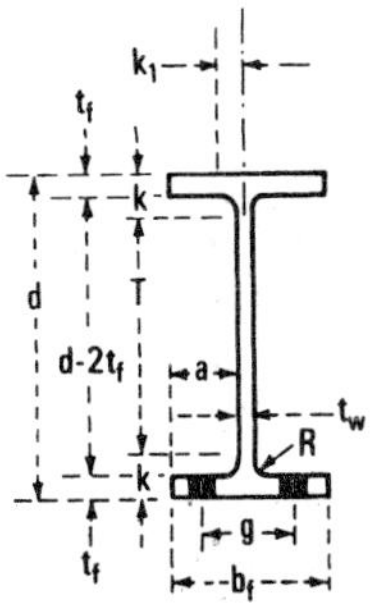

Approximate Dimensions for **Detailing**

Section Number	Weight per Foot	Depth of Section d	Flange Width b_f	Flange Thickness t_f	Web Thickness t_w	Half Web Thickness $\frac{t_w}{2}$	$d-2t_f$	a	T	k	k_1	R	Usual Flange Gage g
	lb	in.	in.	in.	in.	in.	in.	in.	in.	in.	in.	in.	in.
W27 x	178	27 3/4	14 1/8	1 3/16	3/4	3/8	25 7/16	6 5/8	24	1 7/8	1 1/16	0.60	5 1/2
	161	27 5/8	14	1 1/16	11/16	3/8	25 7/16	6 5/8	24	1 13/16	1	0.60	5 1/2
	146	27 3/8	14	1	5/8	5/16	25 7/16	6 5/8	24	1 11/16	1	0.60	5 1/2
W27 x	114	27 1/4	10 1/8	15/16	9/16	5/16	25 7/16	4 3/4	24	1 5/8	15/16	0.60	5 1/2
	102	27 1/8	10	13/16	1/2	1/4	25 7/16	4 3/4	24	1 9/16	15/16	0.60	5 1/2
	94	26 7/8	10	3/4	1/2	1/4	25 7/16	4 3/4	24	1 7/16	15/16	0.60	5 1/2
	84	26 3/4	10	5/8	7/16	1/4	25 7/16	4 3/4	24	1 3/8	15/16	0.60	5 1/2
W24 x	162	25	13	1 1/4	11/16	3/8	22 9/16	6 1/8	21	2	1 1/16	0.50	5 1/2
	146	24 3/4	12 7/8	1 1/16	5/8	5/16	22 9/16	6 1/8	21	1 7/8	1 1/16	0.50	5 1/2
	131	24 1/2	12 7/8	15/16	5/8	5/16	22 9/16	6 1/8	21	1 3/4	1 1/16	0.50	5 1/2
	117	24 1/4	12 3/4	7/8	9/16	5/16	22 9/16	6 1/8	21	1 5/8	1	0.50	5 1/2
	104	24	12 3/4	3/4	1/2	1/4	22 9/16	6 1/8	21	1 1/2	1	0.50	5 1/2
W24 x	94	24 1/4	9 1/8	7/8	1/2	1/4	22 9/16	4 1/4	21	1 5/8	1	0.50	5 1/2
	84	24 1/8	9	3/4	1/2	1/4	22 9/16	4 1/4	21	1 9/16	15/16	0.50	5 1/2
	76	23 7/8	9	11/16	7/16	1/4	22 9/16	4 1/4	21	1 7/16	15/16	0.50	5 1/2
	68	23 3/4	9	9/16	7/16	1/4	22 9/16	4 1/4	21	1 3/8	15/16	0.50	5 1/2
W24 x	62	23 3/4	7	9/16	7/16	1/4	22 9/16	3 1/4	21	1 3/8	15/16	0.50	3 1/2
	55	23 5/8	7	1/2	3/8	3/16	22 9/16	3 1/4	21	1 5/16	15/16	0.50	3 1/2
W21 x	147	22	12 1/2	1 1/8	3/4	3/8	19 3/4	5 7/8	18 1/4	1 7/8	1 1/16	0.50	5 1/2
	132	21 7/8	12 1/2	1 1/16	5/8	5/16	19 3/4	5 7/8	18 1/4	1 13/16	1	0.50	5 1/2
	122	21 5/8	12 3/8	15/16	5/8	5/16	19 3/4	5 7/8	18 1/4	1 11/16	1	0.50	5 1/2
	111	21 1/2	12 3/8	7/8	9/16	5/16	19 3/4	5 7/8	18 1/4	1 5/8	15/16	0.50	5 1/2
	101	21 3/8	12 1/4	13/16	1/2	1/4	19 3/4	5 7/8	18 1/4	1 9/16	15/16	0.50	5 1/2
W21 x	93	21 5/8	8 3/8	15/16	9/16	5/16	19 3/4	3 7/8	18 1/4	1 11/16	1	0.50	5 1/2
	83	21 3/8	8 3/8	13/16	1/2	1/4	19 3/4	3 7/8	18 1/4	1 9/16	15/16	0.50	5 1/2
	73	21 1/4	8 1/4	3/4	7/16	1/4	19 3/4	3 7/8	18 1/4	1 1/2	15/16	0.50	5 1/2
	68	21 1/8	8 1/4	11/16	7/16	1/4	19 3/4	3 7/8	18 1/4	1 7/16	7/8	0.50	5 1/2
	62	21	8 1/4	5/8	3/8	3/16	19 3/4	3 7/8	18 1/4	1 3/8	7/8	0.50	5 1/2
W21 x	57	21	6 1/2	5/8	3/8	3/16	19 3/4	3 1/8	18 1/4	1 3/8	7/8	0.50	3 1/2
	50	20 7/8	6 1/2	9/16	3/8	3/16	19 3/4	3 1/8	18 1/4	1 5/16	7/8	0.50	3 1/2
	44	20 5/8	6 1/2	7/16	3/8	3/16	19 3/4	3 1/8	18 1/4	1 3/16	7/8	0.50	3 1/2

Taken from Bethlehem Steel *Structural Shapes* handbook.

WIDE FLANGE SHAPES

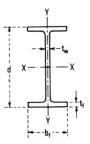

Theoretical Dimensions and Properties for **Designing**

Section Number	Weight per Foot	Area of Section	Depth of Section	Flange		Web Thickness	Axis X-X			Axis Y-Y			r_T
		A	d	Width b_f	Thickness t_f	t_w	I_x	S_x	r_x	I_y	S_y	r_y	
	lb	in.²	in.	in.	in.	in.	in.⁴	in.³	in.	in.⁴	in.³	in.	in.
W18 x 119		35.1	18.97	11.265	1.060	0.655	2190	231	7.90	253	44.9	2.69	3.02
106		31.1	18.73	11.200	0.940	0.590	1910	204	7.84	220	39.4	2.66	3.00
97		28.5	18.59	11.145	0.870	0.535	1750	188	7.82	201	36.1	2.65	2.99
86		25.3	18.39	11.090	0.770	0.480	1530	166	7.77	175	31.6	2.63	2.97
76		22.3	18.21	11.035	0.680	0.425	1330	146	7.73	152	27.6	2.61	2.95
W18 x 71		20.8	18.47	7.635	0.810	0.495	1170	127	7.50	60.3	15.8	1.70	1.98
65		19.1	18.35	7.590	0.750	0.450	1070	117	7.49	54.8	14.4	1.69	1.97
60		17.6	18.24	7.555	0.695	0.415	984	108	7.47	50.1	13.3	1.69	1.96
55		16.2	18.11	7.530	0.630	0.390	890	98.3	7.41	44.9	11.9	1.67	1.95
50		14.7	17.99	7.495	0.570	0.355	800	88.9	7.38	40.1	10.7	1.65	1.94
W18 x 46		13.5	18.06	6.060	0.605	0.360	712	78.8	7.25	22.5	7.43	1.29	1.54
40		11.8	17.90	6.015	0.525	0.315	612	68.4	7.21	19.1	6.35	1.27	1.52
35		10.3	17.70	6.000	0.425	0.300	510	57.6	7.04	15.3	5.12	1.22	1.49
W16 x 100		29.4	16.97	10.425	0.985	0.585	1490	175	7.10	186	35.7	2.52	2.81
89		26.2	16.75	10.365	0.875	0.525	1300	155	7.05	163	31.4	2.49	2.79
77		22.6	16.52	10.295	0.760	0.455	1110	134	7.00	138	26.9	2.47	2.77
67		19.7	16.33	10.235	0.665	0.395	954	117	6.96	119	23.2	2.46	2.75
W16 x 57		16.8	16.43	7.120	0.715	0.430	758	92.2	6.72	43.1	12.1	1.60	1.86
50		14.7	16.26	7.070	0.630	0.380	659	81.0	6.68	37.2	10.5	1.59	1.84
45		13.3	16.13	7.035	0.565	0.345	586	72.7	6.65	32.8	9.34	1.57	1.83
40		11.8	16.01	6.995	0.505	0.305	518	64.7	6.63	28.9	8.25	1.57	1.82
36		10.6	15.86	6.985	0.430	0.295	448	56.5	6.51	24.5	7.00	1.52	1.79
W16 x 31		9.12	15.88	5.525	0.440	0.275	375	47.2	6.41	12.4	4.49	1.17	1.39
26		7.68	15.69	5.500	0.345	0.250	301	38.4	6.26	9.59	3.49	1.12	1.36

All shapes on these pages have parallel-faced flanges.

Taken from Bethlehem Steel *Structural Shapes* handbook.

WIDE FLANGE SHAPES

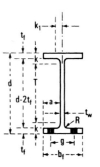

Approximate Dimensions for **Detailing**

Section Number	Weight per Foot	Depth of Section	Flange		Web Thickness	Half Web Thickness	$d-2t_f$	a	T	k	k_1	R	Usual Flange Gage
			Width	Thickness									
		d	b_f	t_f	t_w	$\frac{t_w}{2}$							g
	lb	in.	in.	in.	in.	in.	in.	in.	in.	in.	in.	in.	in.
W18 x	119	19	11¼	1 1/16	5/8	5/16	16⅞	5¼	15½	1¾	15/16	0.40	5½
	106	18¾	11¼	15/16	9/16	5/16	16⅞	5¼	15½	1⅝	15/16	0.40	5½
	97	18⅝	11⅛	⅞	9/16	5/16	16⅞	5¼	15½	1 9/16	⅞	0.40	5½
	86	18⅜	11⅛	¾	½	¼	16⅞	5¼	15½	1 7/16	⅞	0.40	5½
	76	18¼	11	11/16	7/16	¼	16⅞	5¼	15½	1⅜	13/16	0.40	5½
W18 x	71	18½	7⅝	13/16	½	¼	16⅞	3⅝	15½	1½	⅞	0.40	3½
	65	18⅜	7⅝	¾	7/16	¼	16⅞	3⅝	15½	1 7/16	⅞	0.40	3½
	60	18¼	7½	11/16	7/16	¼	16⅞	3⅝	15½	1⅜	13/16	0.40	3½
	55	18⅛	7½	⅝	⅜	3/16	16⅞	3⅝	15½	1 5/16	13/16	0.40	3½
	50	18	7½	9/16	⅜	3/16	16⅞	3⅝	15½	1¼	13/16	0.40	3½
W18 x	46	18	6	⅝	⅜	3/16	16⅞	2⅞	15½	1¼	13/16	0.40	3½
	40	17⅞	6	½	5/16	3/16	16⅞	2⅞	15½	1 3/16	13/16	0.40	3½
	35	17¾	6	7/16	5/16	3/16	16⅞	2⅞	15½	1⅛	¾	0.40	3½
W16 x	100	17	10⅜	1	9/16	5/16	15	4⅞	13⅝	1 11/16	15/16	0.40	5½
	89	16¾	10⅜	⅞	½	¼	15	4⅞	13⅝	1 9/16	⅞	0.40	5½
	77	16½	10¼	¾	7/16	¼	15	4⅞	13⅝	1 7/16	⅞	0.40	5½
	67	16⅜	10¼	11/16	⅜	3/16	15	4⅞	13⅝	1⅜	13/16	0.40	5½
W16 x	57	16⅜	7⅛	11/16	7/16	¼	15	3⅜	13⅝	1⅜	⅞	0.40	3½
	50	16¼	7⅛	⅝	⅜	3/16	15	3⅜	13⅝	1 5/16	13/16	0.40	3½
	45	16⅛	7	9/16	⅜	3/16	15	3⅜	13⅝	1¼	13/16	0.40	3½
	40	16	7	½	5/16	3/16	15	3⅜	13⅝	1 3/16	13/16	0.40	3½
	36	15⅞	7	7/16	5/16	3/16	15	3⅜	13⅝	1⅛	¾	0.40	3½
W16 x	31	15⅞	5½	7/16	¼	⅛	15	2⅝	13⅝	1⅛	¾	0.40	2¾
	26	15¾	5½	⅜	¼	⅛	15	2⅝	13⅝	1 1/16	¾	0.40	2¾

Taken from Bethlehem Steel *Structural Shapes* handbook.

WIDE FLANGE SHAPES

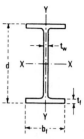

Theoretical Dimensions and Properties for **Designing**

Section Number	Weight per Foot lb	Area of Section A in.²	Depth of Section d in.	Flange Width b_f in.	Flange Thickness t_f in.	Web Thickness t_w in.	Axis X-X I_x in.⁴	Axis X-X S_x in.³	Axis X-X r_x in.	Axis Y-Y I_y in.⁴	Axis Y-Y S_y in.³	Axis Y-Y r_y in.	r_T in.
W14 x	730*	215	22.42	17.890	4.910	3.070	14300	1280	8.17	4720	527	4.69	4.99
	665*	196	21.64	17.650	4.520	2.830	12400	1150	7.98	4170	472	4.62	4.92
	605*	178	20.92	17.415	4.160	2.595	10800	1040	7.80	3680	423	4.55	4.85
	550*	162	20.24	17.200	3.820	2.380	9430	931	7.63	3250	378	4.49	4.79
	500*	147	19.60	17.010	3.500	2.190	8210	838	7.48	2880	339	4.43	4.73
	455*	134	19.02	16.835	3.210	2.015	7190	756	7.33	2560	304	4.38	4.68
W14 x	426	125	18.67	16.695	3.035	1.875	6600	707	7.26	2360	283	4.34	4.64
	398	117	18.29	16.590	2.845	1.770	6000	656	7.16	2170	262	4.31	4.61
	370	109	17.92	16.475	2.660	1.655	5440	607	7.07	1990	241	4.27	4.57
	342	101	17.54	16.360	2.470	1.540	4900	559	6.98	1810	221	4.24	4.54
	311	91.4	17.12	16.230	2.260	1.410	4330	506	6.88	1610	199	4.20	4.50
	283	83.3	16.74	16.110	2.070	1.290	3840	459	6.79	1440	179	4.17	4.46
	257	75.6	16.38	15.995	1.890	1.175	3400	415	6.71	1290	161	4.13	4.43
	233	68.5	16.04	15.890	1.720	1.070	3010	375	6.63	1150	145	4.10	4.40
	211	62.0	15.72	15.800	1.560	0.980	2660	338	6.55	1030	130	4.07	4.37
	193	56.8	15.48	15.710	1.440	0.890	2400	310	6.50	931	119	4.05	4.35
	176	51.8	15.22	15.650	1.310	0.830	2140	281	6.43	838	107	4.02	4.32
	159	46.7	14.98	15.565	1.190	0.745	1900	254	6.38	748	96.2	4.00	4.30
	145	42.7	14.78	15.500	1.090	0.680	1710	232	6.33	677	87.3	3.98	4.28
W14 x	132	38.8	14.66	14.725	1.030	0.645	1530	209	6.28	548	74.5	3.76	4.05
	120	35.3	14.48	14.670	0.940	0.590	1380	190	6.24	495	67.5	3.74	4.04
	109	32.0	14.32	14.605	0.860	0.525	1240	173	6.22	447	61.2	3.73	4.02
	99	29.1	14.16	14.565	0.780	0.485	1110	157	6.17	402	55.2	3.71	4.00
	90	26.5	14.02	14.520	0.710	0.440	999	143	6.14	362	49.9	3.70	3.99
W14 x	82	24.1	14.31	10.130	0.855	0.510	882	123	6.05	148	29.3	2.48	2.74
	74	21.8	14.17	10.070	0.785	0.450	796	112	6.04	134	26.6	2.48	2.72
	68	20.0	14.04	10.035	0.720	0.415	723	103	6.01	121	24.2	2.46	2.71
	61	17.9	13.89	9.995	0.645	0.375	640	92.2	5.98	107	21.5	2.45	2.70
W14 x	53	15.6	13.92	8.060	0.660	0.370	541	77.8	5.89	57.7	14.3	1.92	2.15
	48	14.1	13.79	8.030	0.595	0.340	485	70.3	5.85	51.4	12.8	1.91	2.13
	43	12.6	13.66	7.995	0.530	0.305	428	62.7	5.82	45.2	11.3	1.89	2.12

*These shapes have a 1°-00' (1.75%) flange slope. Flange thicknesses shown are average thicknesses. Properties shown are for a parallel flange section.

Taken from Bethlehem Steel *Structural Shapes* handbook.

WIDE FLANGE SHAPES

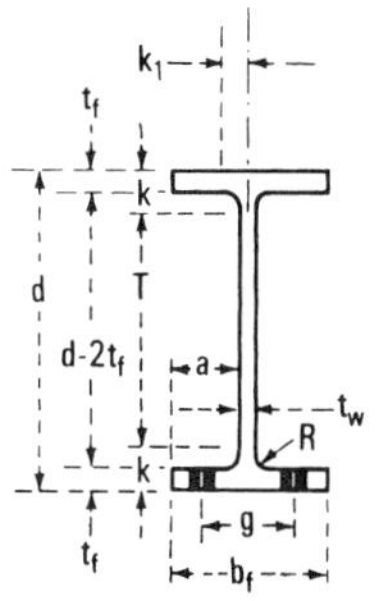

Approximate Dimensions for **Detailing**

Section Number	Weight per Foot	Depth of Section	Flange		Web Thickness	Half Web Thickness	d-2t_f	a	T	k	k_1	R	Usual Flange Gage
			Width	Thickness									
		d	b_f	t_f	t_w	$\frac{t_w}{2}$							g
	lb	in.	in.	in.	in.	in.	in.	in.	in.	in.	in.	in.	in.
W14 x	730	22 3/8	17 7/8	4 15/16	3 1/16	1 9/16	12 5/8	7 3/8	11 1/4	5 9/16	2 3/16	0.60	3-(7 1/2)-3
	665	21 5/8	17 5/8	4 1/2	2 13/16	1 7/16	12 5/8	7 3/8	11 1/4	5 3/16	2 1/16	0.60	3-(7 1/2)-3
	605	20 7/8	17 3/8	4 3/16	2 5/8	1 5/16	12 5/8	7 3/8	11 1/4	4 13/16	1 15/16	0.60	3-(7 1/2)-3
	550	20 1/4	17 1/4	3 13/16	2 3/8	1 3/16	12 5/8	7 3/8	11 1/4	4 1/2	1 13/16	0.60	3-(7 1/2)-3
	500	19 5/8	17	3 1/2	2 3/16	1 1/8	12 5/8	7 3/8	11 1/4	4 3/16	1 3/4	0.60	3-(7 1/2)-3
	455	19	16 7/8	3 3/16	2	1	12 5/8	7 3/8	11 1/4	3 7/8	1 5/8	0.60	3-(7 1/2)-3
W14 x	426	18 5/8	16 3/4	3 1/16	1 7/8	15/16	12 5/8	7 3/8	11 1/4	3 11/16	1 9/16	0.60	3-(5 1/2)-3
	398	18 1/4	16 5/8	2 7/8	1 3/4	7/8	12 5/8	7 3/8	11 1/4	3 1/2	1 1/2	0.60	3-(5 1/2)-3
	370	17 7/8	16 1/2	2 11/16	1 5/8	13/16	12 5/8	7 3/8	11 1/4	3 5/16	1 7/16	0.60	3-(5 1/2)-3
	342	17 1/2	16 3/8	2 1/2	1 9/16	13/16	12 5/8	7 3/8	11 1/4	3 1/8	1 3/8	0.60	3-(5 1/2)-3
	311	17 1/8	16 1/4	2 1/4	1 7/16	3/4	12 5/8	7 3/8	11 1/4	2 15/16	1 5/16	0.60	3-(5 1/2)-3
	283	16 3/4	16 1/8	2 1/16	1 5/16	11/16	12 5/8	7 3/8	11 1/4	2 3/4	1 1/4	0.60	3-(5 1/2)-3
	257	16 3/8	16	1 7/8	1 3/16	5/8	12 5/8	7 3/8	11 1/4	2 9/16	1 3/16	0.60	3-(5 1/2)-3
	233	16	15 7/8	1 3/4	1 1/16	9/16	12 5/8	7 3/8	11 1/4	2 3/8	1 3/16	0.60	3-(5 1/2)-3
	211	15 3/4	15 3/4	1 9/16	1	1/2	12 5/8	7 3/8	11 1/4	2 1/4	1 1/8	0.60	3-(5 1/2)-3
	193	15 1/2	15 3/4	1 7/16	7/8	7/16	12 5/8	7 3/8	11 1/4	2 1/8	1 1/16	0.60	3-(5 1/2)-3
	176	15 1/4	15 5/8	1 5/16	13/16	7/16	12 5/8	7 3/8	11 1/4	2	1 1/16	0.60	3-(5 1/2)-3
	159	15	15 5/8	1 3/16	3/4	3/8	12 5/8	7 3/8	11 1/4	1 7/8	1	0.60	3-(5 1/2)-3
	145	14 3/4	15 1/2	1 1/16	11/16	3/8	12 5/8	7 3/8	11 1/4	1 3/4	1	0.60	3-(5 1/2)-3
W14 x	132	14 5/8	14 3/4	1	5/8	5/16	12 5/8	7	11 1/4	1 11/16	15/16	0.60	5 1/2
	120	14 1/2	14 5/8	15/16	9/16	5/16	12 5/8	7	11 1/4	1 5/8	15/16	0.60	5 1/2
	109	14 3/8	14 5/8	7/8	1/2	1/4	12 5/8	7	11 1/4	1 9/16	7/8	0.60	5 1/2
	99	14 1/8	14 5/8	3/4	1/2	1/4	12 5/8	7	11 1/4	1 7/16	7/8	0.60	5 1/2
	90	14	14 1/2	11/16	7/16	1/4	12 5/8	7	11 1/4	1 3/8	7/8	0.60	5 1/2
W14 x	82	14 1/4	10 1/8	7/8	1/2	1/4	12 5/8	4 3/4	11	1 5/8	1	0.60	5 1/2
	74	14 1/8	10 1/8	13/16	7/16	1/4	12 5/8	4 3/4	11	1 9/16	15/16	0.60	5 1/2
	68	14	10	3/4	7/16	1/4	12 5/8	4 3/4	11	1 1/2	15/16	0.60	5 1/2
	61	13 7/8	10	5/8	3/8	3/16	12 5/8	4 3/4	11	1 7/16	15/16	0.60	5 1/2
W14 x	53	13 7/8	8	11/16	3/8	3/16	12 5/8	3 7/8	11	1 7/16	15/16	0.60	5 1/2
	48	13 3/4	8	5/8	5/16	3/16	12 5/8	3 7/8	11	1 3/8	7/8	0.60	5 1/2
	43	13 5/8	8	1/2	5/16	3/16	12 5/8	3 7/8	11	1 5/16	7/8	0.60	5 1/2

Taken from Bethlehem Steel *Structural Shapes* handbook.

WIDE FLANGE SHAPES

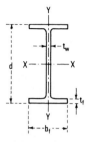

Theoretical Dimensions and Properties for **Designing**

Section Number	Weight per Foot	Area of Section A	Depth of Section d	Flange Width b_f	Flange Thickness t_f	Web Thickness t_w	Axis X-X I_x	Axis X-X S_x	Axis X-X r_x	Axis Y-Y I_y	Axis Y-Y S_y	Axis Y-Y r_y	r_T
	lb	in.²	in.	in.	in.	in.	in.⁴	in.³	in.	in.⁴	in.³	in.	in.
W14 x	38	11.2	14.10	6.770	0.515	0.310	385	54.6	5.88	26.7	7.88	1.55	1.77
	34	10.0	13.98	6.745	0.455	0.285	340	48.6	5.83	23.3	6.91	1.53	1.76
	30	8.85	13.84	6.730	0.385	0.270	291	42.0	5.73	19.6	5.82	1.49	1.74
W14 x	26	7.69	13.91	5.025	0.420	0.255	245	35.3	5.65	8.91	3.54	1.08	1.28
	22	6.49	13.74	5.000	0.335	0.230	199	29.0	5.54	7.00	2.80	1.04	1.25
W12 x	190	55.8	14.38	12.670	1.735	1.060	1890	263	5.82	589	93.0	3.25	3.50
	170	50.0	14.03	12.570	1.560	0.960	1650	235	5.74	517	82.3	3.22	3.47
	152	44.7	13.71	12.480	1.400	0.870	1430	209	5.66	454	72.8	3.19	3.44
	136	39.9	13.41	12.400	1.250	0.790	1240	186	5.58	398	64.2	3.16	3.41
	120	35.3	13.12	12.320	1.105	0.710	1070	163	5.51	345	56.0	3.13	3.38
	106	31.2	12.89	12.220	0.990	0.610	933	145	5.47	301	49.3	3.11	3.36
	96	28.2	12.71	12.160	0.900	0.550	833	131	5.44	270	44.4	3.09	3.34
	87	25.6	12.53	12.125	0.810	0.515	740	118	5.38	241	39.7	3.07	3.32
	79	23.2	12.38	12.080	0.735	0.470	662	107	5.34	216	35.8	3.05	3.31
	72	21.1	12.25	12.040	0.670	0.430	597	97.4	5.31	195	32.4	3.04	3.29
	65	19.1	12.12	12.000	0.605	0.390	533	87.9	5.28	174	29.1	3.02	3.28
W12 x	58	17.0	12.19	10.010	0.640	0.360	475	78.0	5.28	107	21.4	2.51	2.72
	53	15.6	12.06	9.995	0.575	0.345	425	70.6	5.23	95.8	19.2	2.48	2.71
W12 x	50	14.7	12.19	8.080	0.640	0.370	394	64.7	5.18	56.3	13.9	1.96	2.17
	45	13.2	12.06	8.045	0.575	0.335	350	58.1	5.15	50.0	12.4	1.94	2.15
	40	11.8	11.94	8.005	0.515	0.295	310	51.9	5.13	44.1	11.0	1.93	2.14
W12 x	35	10.3	12.50	6.560	0.520	0.300	285	45.6	5.25	24.5	7.47	1.54	1.74
	30	8.79	12.34	6.520	0.440	0.260	238	38.6	5.21	20.3	6.24	1.52	1.73
	26	7.65	12.22	6.490	0.380	0.230	204	33.4	5.17	17.3	5.34	1.51	1.72
W12 x	22	6.48	12.31	4.030	0.425	0.260	156	25.4	4.91	4.66	2.31	0.848	1.02
	19	5.57	12.16	4.005	0.350	0.235	130	21.3	4.82	3.76	1.88	0.822	0.997
	16	4.71	11.99	3.990	0.265	0.220	103	17.1	4.67	2.82	1.41	0.773	0.963
	14	4.16	11.91	3.970	0.225	0.200	88.6	14.9	4.62	2.36	1.19	0.753	0.946

All shapes on these pages have parallel-faced flanges.

Taken from Bethlehem Steel *Structural Shapes* handbook.

WIDE FLANGE SHAPES

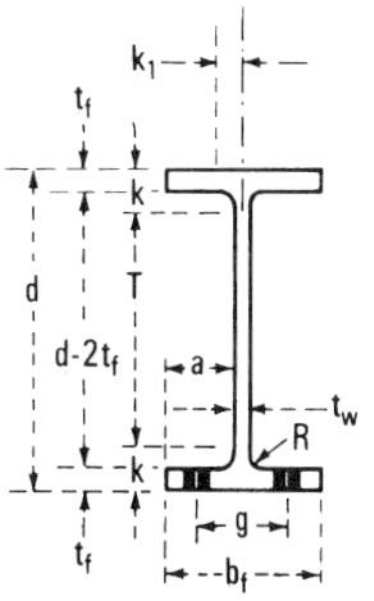

Approximate Dimensions for **Detailing**

Section Number	Weight per Foot	Dept of Section d	Flange Width b_f	Flange Thickness t_f	Web Thickness t_w	Half Web Thickness $\frac{t_w}{2}$	$d-2t_f$	a	T	k	k_1	R	Usual Flange Gage g
	lb	in.	in.	in.	in.	in.	in.	in.	in.	in.	in.	in.	in.
W14 x	**38**	14 1/8	6 3/4	1/2	5/16	3/16	13 1/16	3 1/4	12	1 1/16	5/8	0.40	3 1/2
	34	14	6 3/4	7/16	5/16	3/16	13 1/16	3 1/4	12	1	5/8	0.40	3 1/2
	30	13 7/8	6 3/4	3/8	1/4	1/8	13 1/16	3 1/4	12	15/16	5/8	0.40	3 1/2
W14 x	**26**	13 7/8	5	7/16	1/4	1/8	13 1/16	2 3/8	12	15/16	9/16	0.40	2 3/4
	22	13 3/4	5	5/16	1/4	1/8	13 1/16	2 3/8	12	7/8	9/16	0.40	2 3/4
W12 x	**190**	14 3/8	12 5/8	1 3/4	1 1/16	9/16	10 15/16	5 3/4	9 1/2	2 7/16	1 3/16	0.60	5 1/2
	170	14	12 5/8	1 9/16	15/16	1/2	10 15/16	5 3/4	9 1/2	2 1/4	1 1/8	0.60	5 1/2
	152	13 3/4	12 1/2	1 3/8	7/8	7/16	10 15/16	5 3/4	9 1/2	2 1/8	1 1/16	0.60	5 1/2
	136	13 3/8	12 3/8	1 1/4	13/16	7/16	10 15/16	5 3/4	9 1/2	1 15/16	1	0.60	5 1/2
	120	13 1/8	12 3/8	1 1/8	11/16	3/8	10 15/16	5 3/4	9 1/2	1 13/16	1	0.60	5 1/2
	106	12 7/8	12 1/4	1	5/8	5/16	10 15/16	5 3/4	9 1/2	1 11/16	15/16	0.60	5 1/2
	96	12 3/4	12 1/8	7/8	9/16	5/16	10 15/16	5 3/4	9 1/2	1 5/8	7/8	0.60	5 1/2
	87	12 1/2	12 1/8	13/16	1/2	1/4	10 15/16	5 3/4	9 1/2	1 1/2	7/8	0.60	5 1/2
	79	12 3/8	12 1/8	3/4	1/2	1/4	10 15/16	5 3/4	9 1/2	1 7/16	7/8	0.60	5 1/2
	72	12 1/4	12	11/16	7/16	1/4	10 15/16	5 3/4	9 1/2	1 3/8	7/8	0.60	5 1/2
	65	12 1/8	12	5/8	3/8	3/16	10 15/16	5 3/4	9 1/2	1 5/16	13/16	0.60	5 1/2
W12 x	**58**	12 1/4	10	5/8	3/8	3/16	10 15/16	4 7/8	9 1/2	1 3/8	13/16	0.60	5 1/2
	53	12	10	9/16	3/8	3/16	10 15/16	4 7/8	9 1/2	1 1/4	13/16	0.60	5 1/2
W12 x	**50**	12 1/4	8 1/8	5/8	3/8	3/16	10 15/16	3 7/8	9 1/2	1 3/8	13/16	0.60	5 1/2
	45	12	8	9/16	5/16	3/16	10 15/16	3 7/8	9 1/2	1 1/4	13/16	0.60	5 1/2
	40	12	8	1/2	5/16	3/16	10 15/16	3 7/8	9 1/2	1 1/4	3/4	0.60	5 1/2
W12 x	**35**	12 1/2	6 1/2	1/2	5/16	3/16	11 7/16	3 1/8	10 1/2	1	9/16	0.30	3 1/2
	30	12 3/8	6 1/2	7/16	1/4	1/8	11 7/16	3 1/8	10 1/2	15/16	1/2	0.30	3 1/2
	26	12 1/4	6 1/2	3/8	1/4	1/8	11 7/16	3 1/8	10 1/2	7/8	1/2	0.30	3 1/2
W12 x	**22**	12 1/4	4	7/16	1/4	1/8	11 7/16	1 7/8	10 1/2	7/8	1/2	0.30	2 1/4
	19	12 1/8	4	3/8	1/4	1/8	11 7/16	1 7/8	10 1/2	13/16	1/2	0.30	2 1/4
	16	12	4	1/4	1/4	1/8	11 7/16	1 7/8	10 1/2	3/4	1/2	0.30	2 1/4
	14	11 7/8	4	1/4	3/16	1/8	11 7/16	1 7/8	10 1/2	11/16	1/2	0.30	2 1/4

Taken from Bethlehem Steel *Structural Shapes* handbook.

WIDE FLANGE SHAPES

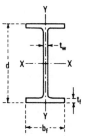

Theoretical Dimensions and Properties for **Designing**

Section Number	Weight per Foot	Area of Section A	Depth of Section d	Flange Width b_f	Flange Thickness t_f	Web Thickness t_w	Axis X-X I_x	Axis X-X S_x	Axis X-X r_x	Axis Y-Y I_y	Axis Y-Y S_y	Axis Y-Y r_y	r_T
	lb	in.²	in.	in.	in.	in.	in.⁴	in.³	in.	in.⁴	in.³	in.	in.
W10 x	112	32.9	11.36	10.415	1.250	0.755	716	126	4.66	236	45.3	2.68	2.88
	100	29.4	11.10	10.340	1.120	0.680	623	112	4.60	207	40.0	2.65	2.85
	88	25.9	10.84	10.265	0.990	0.605	534	98.5	4.54	179	34.8	2.63	2.83
	77	22.6	10.60	10.190	0.870	0.530	455	85.9	4.49	154	30.1	2.60	2.80
	68	20.0	10.40	10.130	0.770	0.470	394	75.7	4.44	134	26.4	2.59	2.79
	60	17.6	10.22	10.080	0.680	0.420	341	66.7	4.39	116	23.0	2.57	2.77
	54	15.8	10.09	10.030	0.615	0.370	303	60.0	4.37	103	20.6	2.56	2.75
	49	14.4	9.98	10.000	0.560	0.340	272	54.6	4.35	93.4	18.7	2.54	2.74
W10 x	45	13.3	10.10	8.020	0.620	0.350	248	49.1	4.33	53.4	13.3	2.01	2.18
	39	11.5	9.92	7.985	0.530	0.315	209	42.1	4.27	45.0	11.3	1.98	2.16
	33	9.71	9.73	7.960	0.435	0.290	170	35.0	4.19	36.6	9.20	1.94	2.14
W10 x	30	8.84	10.47	5.810	0.510	0.300	170	32.4	4.38	16.7	5.75	1.37	1.55
	26	7.61	10.33	5.770	0.440	0.260	144	27.9	4.35	14.1	4.89	1.36	1.54
	22	6.49	10.17	5.750	0.360	0.240	118	23.2	4.27	11.4	3.97	1.33	1.51
W10 x	19	5.62	10.24	4.020	0.395	0.250	96.3	18.8	4.14	4.29	2.14	0.874	1.03
	17	4.99	10.11	4.010	0.330	0.240	81.9	16.2	4.05	3.56	1.78	0.845	1.01
	15	4.41	9.99	4.000	0.270	0.230	68.9	13.8	3.95	2.89	1.45	0.810	0.987
	12	3.54	9.87	3.960	0.210	0.190	53.8	10.9	3.90	2.18	1.10	0.785	0.965
W8 x	67	19.7	9.00	8.280	0.935	0.570	272	60.4	3.72	88.6	21.4	2.12	2.28
	58	17.1	8.75	8.220	0.810	0.510	228	52.0	3.65	75.1	18.3	2.10	2.26
	48	14.1	8.50	8.110	0.685	0.400	184	43.3	3.61	60.9	15.0	2.08	2.23
	40	11.7	8.25	8.070	0.560	0.360	146	35.5	3.53	49.1	12.2	2.04	2.21
	35	10.3	8.12	8.020	0.495	0.310	127	31.2	3.51	42.6	10.6	2.03	2.20
	31	9.13	8.00	7.995	0.435	0.285	110	27.5	3.47	37.1	9.27	2.02	2.18
W8 x	28	8.25	8.06	6.535	0.465	0.285	98.0	24.3	3.45	21.7	6.63	1.62	1.77
	24	7.08	7.93	6.495	0.400	0.245	82.8	20.9	3.42	18.3	5.63	1.61	1.76
W8 x	21	6.16	8.28	5.270	0.400	0.250	75.3	18.2	3.49	9.77	3.71	1.26	1.41
	18	5.26	8.14	5.250	0.330	0.230	61.9	15.2	3.43	7.97	3.04	1.23	1.39
W8 x	15	4.44	8.11	4.015	0.315	0.245	48.0	11.8	3.29	3.41	1.70	0.876	1.03
	13	3.84	7.99	4.000	0.255	0.230	39.6	9.91	3.21	2.73	1.37	0.843	1.01
	10	2.96	7.89	3.940	0.205	0.170	30.8	7.81	3.22	2.09	1.06	0.841	0.994

All shapes on these pages have parallel-faced flanges.

Taken from Bethlehem Steel *Structural Shapes* handbook.

WIDE FLANGE SHAPES

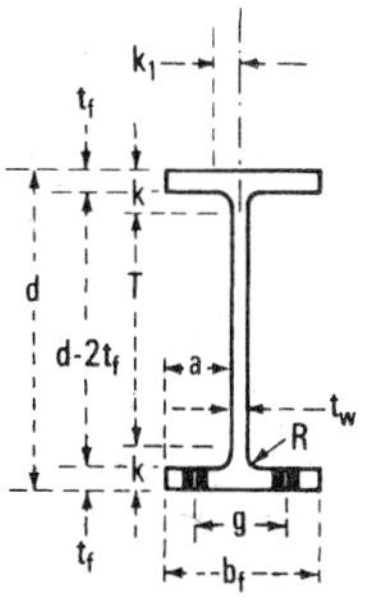

Approximate Dimensions for **Detailing**

Section Number	Weight per Foot	Depth of Section d	Flange Width b_f	Flange Thickness t_f	Web Thickness t_w	Half Web Thickness $\frac{t_w}{2}$	$d-2t_f$	a	T	k	k_1	R	Usual Flange Gage g
	lb	in.	in.	in.	in.	in.	in.	in.	in.	in.	in.	in.	in.
W10 x	112	11⅜	10⅜	1¼	¾	⅜	8⅞	4⅞	7⅝	1⅞	15/16	0.50	5½
	100	11⅛	10⅜	1⅛	11/16	⅜	8⅞	4⅞	7⅝	1¾	⅞	0.50	5½
	88	10⅞	10¼	1	⅝	5/16	8⅞	4⅞	7⅝	1⅝	13/16	0.50	5½
	77	10⅝	10¼	⅞	½	¼	8⅞	4⅞	7⅝	1½	13/16	0.50	5½
	68	10⅜	10⅛	¾	½	¼	8⅞	4⅞	7⅝	1⅜	¾	0.50	5½
	60	10¼	10⅛	11/16	7/16	¼	8⅞	4⅞	7⅝	1 5/16	¾	0.50	5½
	54	10⅛	10	⅝	⅜	3/16	8⅞	4⅞	7⅝	1¼	11/16	0.50	5½
	49	10	10	9/16	5/16	3/16	8⅞	4⅞	7⅝	1 3/16	11/16	0.50	5½
W10 x	45	10⅛	8	⅝	⅜	3/16	8⅞	3⅞	7⅝	1¼	11/16	0.50	5½
	39	9⅞	8	½	5/16	3/16	8⅞	3⅞	7⅝	1⅛	11/16	0.50	5½
	33	9¾	8	7/16	5/16	3/16	8⅞	3⅞	7⅝	1 1/16	11/16	0.50	5½
W10 x	30	10½	5¾	½	5/16	3/16	9 7/16	2¾	8⅝	15/16	½	0.30	2¾
	26	10⅜	5¾	7/16	¼	⅛	9 7/16	2¾	8⅝	⅞	½	0.30	2¾
	22	10⅛	5¾	⅜	¼	⅛	9 7/16	2¾	8⅝	¾	½	0.30	2¾
W10 x	19	10¼	4	⅜	¼	⅛	9 7/16	1⅞	8⅝	13/16	½	0.30	2¼
	17	10⅛	4	5/16	¼	⅛	9 7/16	1⅞	8⅝	¾	½	0.30	2¼
	15	10	4	¼	¼	⅛	9 7/16	1⅞	8⅝	11/16	7/16	0.30	2¼
	12	9⅞	4	3/16	3/16	⅛	9 7/16	1⅞	8⅝	⅝	7/16	0.30	2¼
W8 x	67	9	8¼	15/16	9/16	5/16	7⅛	3⅞	6⅛	1 7/16	11/16	0.40	5½
	58	8¾	8¼	13/16	½	¼	7⅛	3⅞	6⅛	1 5/16	11/16	0.40	5½
	48	8½	8⅛	11/16	⅜	3/16	7⅛	3⅞	6⅛	1 3/16	⅝	0.40	5½
	40	8¼	8⅛	9/16	⅜	3/16	7⅛	3⅞	6⅛	1 1/16	⅝	0.40	5½
	35	8⅛	8	½	5/16	3/16	7⅛	3⅞	6⅛	1	9/16	0.40	5½
	31	8	8	7/16	5/16	3/16	7⅛	3⅞	6⅛	15/16	9/16	0.40	5½
W8 x	28	8	6½	7/16	5/16	3/16	7⅛	3⅛	6⅛	15/16	9/16	0.40	3½
	24	7⅞	6½	⅜	¼	⅛	7⅛	3⅛	6⅛	⅞	9/16	0.40	3½
W8 x	21	8¼	5¼	⅜	¼	⅛	7½	2½	6⅝	13/16	½	0.30	2¾
	18	8⅛	5¼	5/16	¼	⅛	7½	2½	6⅝	¾	7/16	0.30	2¾
W8 x	15	8⅛	4	5/16	¼	⅛	7½	1⅞	6⅝	¾	½	0.30	2¼
	13	8	4	¼	¼	⅛	7½	1⅞	6⅝	11/16	7/16	0.30	2¼
	10	7⅞	4	3/16	3/16	⅛	7½	1⅞	6⅝	⅝	7/16	0.30	2¼

Taken from Bethlehem Steel *Structural Shapes* handbook.

WIDE FLANGE SHAPES

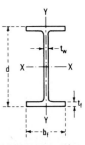

Theoretical Dimensions and Properties for **Designing**

Section Number	Weight per Foot	Area of Section	Depth of Section	Flange		Web Thickness	Axis X-X			Axis Y-Y			r_T
				Width	Thickness								
		A	d	b_f	t_f	t_w	I_x	S_x	r_x	I_y	S_y	r_y	
	lb	in.²	in.	in.	in.	in.	in.⁴	in.³	in.	in.⁴	in.³	in.	in.
W6 x	25	7.34	6.38	6.080	0.455	0.320	53.4	16.7	2.70	17.1	5.61	1.52	1.66
	20	5.87	6.20	6.020	0.365	0.260	41.4	13.4	2.66	13.3	4.41	1.50	1.64
	15	4.43	5.99	5.990	0.260	0.230	29.1	9.72	2.56	9.32	3.11	1.45	1.61
W6 x	16	4.74	6.28	4.030	0.405	0.260	32.1	10.2	2.60	4.43	2.20	0.967	1.08
	12	3.55	6.03	4.000	0.280	0.230	22.1	7.31	2.49	2.99	1.50	0.918	1.05
	9	2.68	5.90	3.940	0.215	0.170	16.4	5.56	2.47	2.20	1.11	0.905	1.03
W5 x	19	5.54	5.15	5.030	0.430	0.270	26.2	10.2	2.17	9.13	3.63	1.28	1.38
	16	4.68	5.01	5.000	0.360	0.240	21.3	8.51	2.13	7.51	3.00	1.27	1.37
†W4 x	13	3.83	4.16	4.060	0.345	0.280	11.3	5.46	1.72	3.86	1.90	1.00	1.10

MISCELLANEOUS SHAPE

Theoretical Dimensions and Properties for **Designing**

Section Number	Weight per Foot	Area of Section	Depth of Section	Flange		Web Thickness	Axis X-X			Axis Y-Y			r_T
				Width	Thickness								
		A	d	b_f	t_f	t_w	I_x	S_x	r_x	I_y	S_y	r_y	
	lb	in.²	in.	in.	in.	in.	in.⁴	in.³	in.	in.⁴	in.³	in.	in.
†M5 x	18.9	5.55	5.00	5.003	0.416	0.316	24.1	9.63	2.08	7.86	3.14	1.19	1.32

†W4 x 13 and M5 x 18.9 have flange slopes of 2.0 and 7.4 pct respectively. Flange thickness shown for these sections are average thicknesses. Properties are the same as if flanges were parallel.

All other shapes on these pages have parallel-faced flanges.

Taken from Bethlehem Steel *Structural Shapes* handbook.

WIDE FLANGE SHAPES

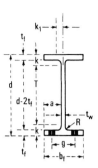

Approximate Dimensions for **Detailing**

Section Number	Weight per Foot	Depth of Section	Flange		Web Thickness	Half Web Thickness	$d-2t_f$	a	T	k	k_1	R	Usual Flange Gage
			Width	Thickness									
		d	b_f	t_f	t_w	$\frac{t_w}{2}$							g
	lb	in.	in.	in.	in.	in.	in.	in.	in.	in.	in.	in.	in.
W6 x	25	6⅜	6⅛	7/16	5/16	3/16	5½	2⅞	4¾	13/16	7/16	0.25	3½
	20	6¼	6	⅜	¼	⅛	5½	2⅞	4¾	¾	7/16	0.25	3½
	15	6	6	¼	¼	⅛	5½	2⅞	4¾	⅝	⅜	0.25	3½
W6 x	16	6¼	4	⅜	¼	⅛	5½	1⅞	4¾	¾	7/16	0.25	2¼
	12	6	4	¼	¼	⅛	5½	1⅞	4¾	⅝	⅜	0.25	2¼
	9	5⅞	4	3/16	3/16	⅛	5½	1⅞	4¾	9/16	⅜	0.25	2¼
W5 x	19	5⅛	5	7/16	¼	⅛	4 5/16	2⅜	3½	13/16	7/16	0.30	2¾
	16	5	5	⅜	¼	⅛	4 5/16	2⅜	3½	¾	7/16	0.30	2¾
W4 x	13	4⅛	4	⅜	¼	⅛	3½	1⅞	2¾	11/16	7/16	0.25	2¼

MISCELLANEOUS SHAPE

Theoretical Dimensions and Properties for **Detailing**

Section Number	Weight per Foot	Depth of Section	Flange		Web Thickness	Half Web Thickness	$d-2t_f$	a	T	k	k_1	R	Usual Flange Gage
			Width	Thickness									
		d	b_f	t_f	t_w	$\frac{t_w}{2}$							g
	lb	in.	in.	in.	in.	in.	in.	in.	in.	in.	in.	in.	in.
M5 x	18.9	5	5	7/16	5/16	3/16	4 3/16	2⅜	3¼	⅞	½	0.313	2¾

Taken from Bethlehem Steel *Structural Shapes* handbook.

H-PILES

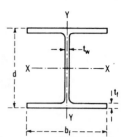

Theoretical Dimensions and Properties for Designing

Section Number	Weight per Foot	Area of Section	Depth of Section	Flange		Web Thickness	Axis X-X			Axis Y-Y			r_T
		A	d	Width b_f	Thickness t_f	t_w	I_x	S_x	r_x	I_y	S_y	r_y	
	lb	in.^{2}	in.	in.	in.	in.	in.^{4}	in.^{3}	in.	in.^{4}	in.^{3}	in.	in.
HP14 x	117	34.4	14.21	14.885	0.805	0.805	1220	172	5.96	443	59.5	3.59	4.00
	102	30.0	14.01	14.785	0.705	0.705	1050	150	5.92	380	51.4	3.56	3.97
	89	26.1	13.83	14.695	0.615	0.615	904	131	5.88	326	44.3	3.53	3.94
	73	21.4	13.61	14.585	0.505	0.505	729	107	5.84	261	35.8	3.49	3.90
HP12 x	74	21.8	12.13	12.215	0.610	0.605	569	93.8	5.11	186	30.4	2.92	3.26
	63	18.4	11.94	12.125	0.515	0.515	472	79.1	5.06	153	25.3	2.88	3.23
	53	15.5	11.78	12.045	0.435	0.435	393	66.8	5.03	127	21.1	2.86	3.20
HP10 x	57	16.8	9.99	10.225	0.565	0.565	294	58.8	4.18	101	19.7	2.45	2.74
	42	12.4	9.70	10.075	0.420	0.415	210	43.4	4.13	71.7	14.2	2.41	2.69
HP8 x	36	10.6	8.02	8.155	0.445	0.445	119	29.8	3.36	40.3	9.88	1.95	2.18

H-Piles have parallel-faced flanges.

HEAVY DUTY SHEET PILING

Theoretical Dimensions and Properties

Section Number	Area of Section	Nominal Depth D	Nominal Width W	Thickness		Weight		Moment of Inertia	Section Modulus	
				Flange t_f	Web t_w	per lin. ft of bar	per sq. ft of wall		Single Section	per lin. ft of wall
	in.^{2}	in.	in.	in.	in.	in.	lb	in.^{4}	in.^{3}	in.^{3}
PZ22	11.86	9.0	22.0	.375	.375	40.3	22	154.7	33.1	18.1
PZ27	11.91	12.0	18.0	.375	.375	40.5	27	276.3	45.3	30.2
PZ35	19.41	14.9	22.64	.60	.50	66.0	35	681.5	91.4	48.5
PZ40	19.30	16.1	19.69	.60	.50	65.6	40	805.4	99.6	60.7
PSA23	8.99	1.3	16.0	—	.375	30.7	23	5.5	3.2	2.4
PS27.5	13.27	—	19.69	—	.40	45.1	27.5	5.3	3.3	2.0
PS31	14.96	—	19.69	—	.50	50.9	31	5.3	3.3	2.0

*Sheet piling is available in ASTM A328, A572 Grades 50 and 60, and A690.
For complete details on sheet piling, refer to the Bethlehem Steel Sheet Piling Booklet NO. 2001.

Taken from Bethlehem Steel *Structural Shapes* handbook.

H-PILES

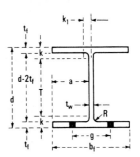

Approximate Dimensions for **Detailing**

Section Number	Weight per Foot	Depth of Section d	Flange Width b_f	Flange Thickness t_f	Web Thickness t_w	Half Web Thickness $\frac{t_w}{2}$	$d-2t_f$	a	T	k	k_1	R	Usual Flange Gage g
	lb	in.	in.	in.	in.	in.	in.	in.	in.	in.	in.	in.	in.
HP14 x	117	14 1/4	14 7/8	13/16	13/16	7/16	12 5/8	7	11 1/4	1 1/2	1 1/16	0.60	5 1/2
	102	14	14 3/4	11/16	11/16	3/8	12 5/8	7	11 1/4	1 3/8	1	0.60	5 1/2
	89	13 7/8	14 3/4	5/8	5/8	5/16	12 5/8	7	11 1/4	1 5/16	15/16	0.60	5 1/2
	73	13 5/8	14 5/8	1/2	1/2	1/4	12 5/8	7	11 1/4	1 3/16	7/8	0.60	5 1/2
HP12 x	74	12 1/8	12 1/4	5/8	5/8	5/16	10 15/16	5 3/4	9 1/2	1 5/16	15/16	0.60	5 1/2
	63	12	12 1/8	1/2	1/2	1/4	10 15/16	5 3/4	9 1/2	1 1/4	7/8	0.60	5 1/2
	53	11 3/4	12	7/16	7/16	1/4	10 15/16	5 3/4	9 1/2	1 1/8	7/8	0.60	5 1/2
HP10 x	57	10	10 1/4	9/16	9/16	5/16	8 7/8	4 7/8	7 5/8	1 3/16	13/16	0.50	5 1/2
	42	9 3/4	10 1/8	7/16	7/16	1/4	8 7/8	4 7/8	7 5/8	1 1/16	3/4	0.50	5 1/2
HP8 x	36	8	8 1/8	7/16	7/16	1/4	7 1/8	3 7/8	6 1/8	15/16	5/8	0.40	5 1/2

HEAVY DUTY SHEET PILING

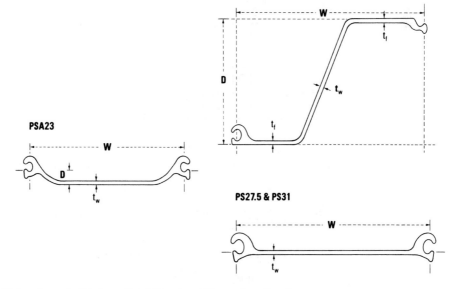

Taken from Bethlehem Steel *Structural Shapes* handbook.

STRUCTURAL TEES (Cut from W Shapes)

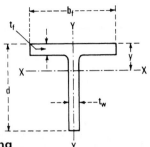

Theoretical Dimensions and Properties for **Designing**

Section Number	Weight per Foot	Area of Section	Depth of Section	Flange		Stem Thickness	Axis X-X				Axis Y-Y		
				Width	Thickness								
		A	d	b_f	t_f	t_w	I_x	S_x	r_x	y	I_y	S_y	r_y
	lb	in.²	in.	in.	in.	in.	in.⁴	in.³	in.	in.	in.⁴	in.³	in.
WT18 x 150		44.1	18.370	16.655	1.680	0.945	1230	86.1	5.27	4.13	648	77.8	3.83
140		41.2	18.260	16.595	1.570	0.885	1140	80.0	5.25	4.07	599	72.2	3.81
130		38.2	18.130	16.550	1.440	0.840	1060	75.1	5.26	4.05	545	65.9	3.78
122.5		36.0	18.040	16.510	1.350	0.800	995	71.0	5.26	4.03	507	61.4	3.75
115		33.8	17.950	16.470	1.260	0.760	934	67.0	5.25	4.01	470	57.1	3.73
WT18 x 105		30.9	18.345	12.180	1.360	0.830	985	73.1	5.65	4.87	206	33.8	2.58
97		28.5	18.245	12.115	1.260	0.765	901	67.0	5.62	4.80	187	30.9	2.56
91		26.8	18.165	12.075	1.180	0.725	845	63.1	5.62	4.77	174	28.8	2.55
85		25.0	18.085	12.030	1.100	0.680	786	58.9	5.61	4.73	160	26.6	2.53
80		23.5	18.005	12.000	1.020	0.650	740	55.8	5.61	4.74	147	24.6	2.50
75		22.1	17.925	11.975	0.940	0.625	698	53.1	5.62	4.78	135	22.5	2.47
67.5		19.9	17.775	11.950	0.790	0.600	636	49.7	5.66	4.96	113	18.9	2.38
WT16.5 x 120.5		35.4	17.090	15.860	1.400	0.830	871	65.8	4.96	3.85	466	58.8	3.63
110.5		32.5	16.965	15.805	1.275	0.775	799	60.8	4.96	3.81	420	53.2	3.59
100.5		29.5	16.840	15.745	1.150	0.715	725	55.5	4.95	3.78	375	47.6	3.56
WT16.5 x 76		22.4	16.745	11.565	1.055	0.635	592	47.4	5.14	4.26	136	23.6	2.47
70.5		20.8	16.650	11.535	0.960	0.605	552	44.7	5.15	4.29	123	21.3	2.43
65		19.2	16.545	11.510	0.855	0.580	513	42.1	5.18	4.36	109	18.9	2.39
59		17.3	16.430	11.480	0.740	0.550	469	39.2	5.20	4.47	93.6	16.3	2.32
WT15 x 105.5		31.0	15.470	15.105	1.315	0.775	610	50.5	4.43	3.40	378	50.1	3.49
95.5		28.1	15.340	15.040	1.185	0.710	549	45.7	4.42	3.35	336	44.7	3.46
86.5		25.4	15.220	14.985	1.065	0.655	497	41.7	4.42	3.31	299	39.9	3.43
WT15 x 66		19.4	15.155	10.545	1.000	0.615	421	37.4	4.66	3.90	98.0	18.6	2.25
62		18.2	15.085	10.515	0.930	0.585	396	35.3	4.66	3.90	90.4	17.2	2.23
58		17.1	15.005	10.495	0.850	0.565	373	33.7	4.67	3.94	82.1	15.7	2.19
54		15.9	14.915	10.475	0.760	0.545	349	32.0	4.69	4.01	73.0	13.9	2.15
49.5		14.5	14.825	10.450	0.670	0.520	322	30.0	4.71	4.09	63.9	12.2	2.10
WT13.5 x 89		26.1	13.905	14.085	1.190	0.725	414	38.2	3.98	3.05	278	39.4	3.26
80.5		23.7	13.795	14.020	1.080	0.660	372	34.4	3.96	2.99	248	35.4	3.24
73		21.5	13.690	13.965	0.975	0.605	336	31.2	3.95	2.95	222	31.7	3.21
WT13.5 x 57		16.8	13.645	10.070	0.930	0.570	289	28.3	4.15	3.42	79.4	15.8	2.18
51		15.0	13.545	10.015	0.830	0.515	258	25.3	4.14	3.37	69.6	13.9	2.15
47		13.8	13.460	9.990	0.745	0.490	239	23.8	4.16	3.41	62.0	12.4	2.12
42		12.4	13.355	9.960	0.640	0.460	216	21.9	4.18	3.48	52.8	10.6	2.07

Properties shown in this table are for the full center split section.

Taken from Bethlehem Steel *Structural Shapes* handbook.

STRUCTURAL TEES (Cut from W Shapes)

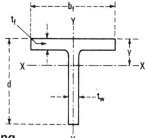

Theoretical Dimensions and Properties for **Designing**

Section Number	Weight per Foot	Area of Section	Depth of Section	Flange		Stem Thickness	Axis X-X				Axis Y-Y		
				Width	Thickness								
		A	d	b_f	t_f	t_w	I_x	S_x	r_x	y	I_y	S_y	r_y
	lb	in.²	in.	in.	in.	in.	in.⁴	in.³	in.	in.	in.⁴	in.³	in.
WT12 x 81		23.9	12.500	12.955	1.220	0.705	293	29.9	3.50	2.70	221	34.2	3.05
73		21.5	12.370	12.900	1.090	0.650	264	27.2	3.50	2.66	195	30.3	3.01
65.5		19.3	12.240	12.855	0.960	0.605	238	24.8	3.52	2.65	170	26.5	2.97
58.5		17.2	12.130	12.800	0.850	0.550	212	22.3	3.51	2.62	149	23.2	2.94
52		15.3	12.030	12.750	0.750	0.500	189	20.0	3.51	2.59	130	20.3	2.91
WT12 x 47		13.8	12.155	9.065	0.875	0.515	186	20.3	3.67	2.99	54.5	12.0	1.98
42		12.4	12.050	9.020	0.770	0.470	166	18.3	3.67	2.97	47.2	10.5	1.95
38		11.2	11.960	8.990	0.680	0.440	151	16.9	3.68	3.00	41.3	9.18	1.92
34		10.0	11.865	8.965	0.585	0.415	137	15.6	3.70	3.06	35.2	7.85	1.87
WT12 x 31		9.11	11.870	7.040	0.590	0.430	131	15.6	3.79	3.46	17.2	4.90	1.38
27.5		8.10	11.785	7.005	0.505	0.395	117	14.1	3.80	3.50	14.5	4.15	1.34
WT10.5 x 73.5		21.6	11.030	12.510	1.150	0.720	204	23.7	3.08	2.39	188	30.0	2.95
66		19.4	10.915	12.440	1.035	0.650	181	21.1	3.06	2.33	166	26.7	2.93
61		17.9	10.840	12.390	0.960	0.600	166	19.3	3.04	2.28	152	24.6	2.92
55.5		16.3	10.755	12.340	0.875	0.550	150	17.5	3.03	2.23	137	22.2	2.90
50.5		14.9	10.680	12.290	0.800	0.500	135	15.8	3.01	2.18	124	20.2	2.89
WT10.5 x 46.5		13.7	10.810	8.420	0.930	0.580	144	17.9	3.25	2.74	46.4	11.0	1.84
41.5		12.2	10.715	8.355	0.835	0.515	127	15.7	3.22	2.66	40.7	9.75	1.83
36.5		10.7	10.620	8.295	0.740	0.455	110	13.8	3.21	2.60	35.3	8.51	1.81
34		10.0	10.565	8.270	0.685	0.430	103	12.9	3.20	2.59	32.4	7.83	1.80
31		9.13	10.495	8.240	0.615	0.400	93.8	11.9	3.21	2.58	28.7	6.97	1.77
WT10.5 x 28.5		8.37	10.530	6.555	0.650	0.405	90.4	11.8	3.29	2.85	15.3	4.67	1.35
25		7.36	10.415	6.530	0.535	0.380	80.3	10.7	3.30	2.93	12.5	3.82	1.30
22		6.49	10.330	6.500	0.450	0.350	71.1	9.68	3.31	2.98	10.3	3.18	1.26
WT9 x 59.5		17.5	9.485	11.265	1.060	0.655	119	15.9	2.60	2.03	126	22.5	2.69
53		15.6	9.365	11.200	0.940	0.590	104	14.1	2.59	1.97	110	19.7	2.66
48.5		14.3	9.295	11.145	0.870	0.535	93.8	12.7	2.56	1.91	100	18.0	2.65
43		12.7	9.195	11.090	0.770	0.480	82.4	11.2	2.55	1.86	87.6	15.8	2.63
38		11.2	9.105	11.035	0.680	0.425	71.8	9.83	2.54	1.80	76.2	13.8	2.61
WT9 x 35.5		10.4	9.235	7.635	0.810	0.495	78.2	11.2	2.74	2.26	30.1	7.89	1.70
32.5		9.55	9.175	7.590	0.750	0.450	70.7	10.1	2.72	2.20	27.4	7.22	1.69
30		8.82	9.120	7.555	0.695	0.415	64.7	9.29	2.71	2.16	25.0	6.63	1.69
27.5		8.10	9.055	7.530	0.630	0.390	59.5	8.63	2.71	2.16	22.5	5.97	1.67
25		7.33	8.995	7.495	0.570	0.355	53.5	7.79	2.70	2.12	20.0	5.35	1.65
WT9 x 23		6.77	9.030	6.060	0.605	0.360	52.1	7.77	2.77	2.33	11.3	3.72	1.29
20		5.88	8.950	6.015	0.525	0.315	44.8	6.73	2.76	2.29	9.55	3.17	1.27
17.5		5.15	8.850	6.000	0.425	0.300	40.1	6.21	2.79	2.39	7.67	2.56	1.22

Properties shown in this table are for the full center split section.

Taken from Bethlehem Steel *Structural Shapes* handbook.

STRUCTURAL TEES (Cut from W Shapes)

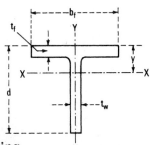

Theoretical Dimensions and Properties for **Designing**

Section Number	Weight per Foot	Area of Section	Depth of Section	Flange		Stem Thickness	Axis X-X				Axis Y-Y		
				Width	Thickness								
		A	d	b_f	t_f	t_w	I_x	S_x	r_x	y	I_y	S_y	r_y
	lb	in.²	in.	in.	in.	in.	in.⁴	in.³	in.	in.	in.⁴	in.³	in.
WT8 x	50	14.7	8.485	10.425	0.985	0.585	76.8	11.4	2.28	1.76	93.1	17.9	2.52
	44.5	13.1	8.375	10.365	0.875	0.525	67.2	10.1	2.27	1.70	81.3	15.7	2.49
	38.5	11.3	8.260	10.295	0.760	0.455	56.9	8.59	2.24	1.63	69.2	13.4	2.47
	33.5	9.84	8.165	10.235	0.665	0.395	48.6	7.36	2.22	1.56	59.5	11.6	2.46
WT8 x	28.5	8.38	8.215	7.120	0.715	0.430	48.7	7.77	2.41	1.94	21.6	6.06	1.60
	25	7.37	8.130	7.070	0.630	0.380	42.3	6.78	2.40	1.89	18.6	5.26	1.59
	22.5	6.63	8.065	7.035	0.565	0.345	37.8	6.10	2.39	1.86	16.4	4.67	1.57
	20	5.89	8.005	6.995	0.505	0.305	33.1	5.35	2.37	1.81	14.4	4.12	1.57
	18	5.28	7.930	6.985	0.430	0.295	30.6	5.05	2.41	1.88	12.2	3.50	1.52
WT8 x	15.5	4.56	7.940	5.525	0.440	0.275	27.4	4.64	2.45	2.02	6.20	2.24	1.17
	13	3.84	7.845	5.500	0.345	0.250	23.5	4.09	2.47	2.09	4.80	1.74	1.12
WT7 x	365	107	11.210	17.890	4.910	3.070	739	95.4	2.62	3.47	2360	264	4.69
	332.5	97.8	10.820	17.650	4.520	2.830	622	82.1	2.52	3.25	2080	236	4.62
	302.5	88.9	10.460	17.415	4.160	2.595	524	70.6	2.43	3.05	1840	211	4.55
	275	80.9	10.120	17.200	3.820	2.380	442	60.9	2.34	2.85	1630	189	4.49
	250	73.5	9.800	17.010	3.500	2.190	375	52.7	2.26	2.67	1440	169	4.43
	227.5	66.9	9.510	16.835	3.210	2.015	321	45.9	2.19	2.51	1280	152	4.38
WT7 x	213	62.6	9.335	16.695	3.035	1.875	287	41.4	2.14	2.40	1180	141	4.34
	199	58.5	9.145	16.590	2.845	1.770	257	37.6	2.10	2.30	1090	131	4.31
	185	54.4	8.960	16.475	2.660	1.655	229	33.9	2.05	2.19	994	121	4.27
	171	50.3	8.770	16.360	2.470	1.540	203	30.4	2.01	2.09	903	110	4.24
	155.5	45.7	8.560	16.230	2.260	1.410	176	26.7	1.96	1.97	807	99.4	4.20
	141.5	41.6	8.370	16.110	2.070	1.290	153	23.5	1.92	1.86	722	89.7	4.17
	128.5	37.8	8.190	15.995	1.890	1.175	133	20.7	1.88	1.75	645	80.7	4.13
	116.5	34.2	8.020	15.890	1.720	1.070	116	18.2	1.84	1.65	576	72.5	4.10
	105.5	31.0	7.860	15.800	1.560	0.098	102	16.2	1.81	1.57	513	65.0	4.07
	96.5	28.4	7.740	15.710	1.440	0.890	89.8	14.4	1.78	1.49	466	59.3	4.05
	88	25.9	7.610	15.650	1.310	0.830	80.5	13.0	1.76	1.43	419	53.5	4.02
	79.5	23.4	7.490	15.565	1.190	0.745	70.2	11.4	1.73	1.35	374	48.1	4.00
	72.5	21.3	7.390	15.500	1.090	0.680	62.5	10.2	1.71	1.29	338	43.7	3.98

Properties shown in this table are for the full center split section.

Taken from Bethlehem Steel *Structural Shapes* handbook.

STRUCTURAL TEES (Cut from W Shapes)

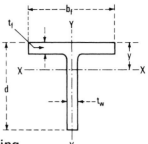

Theoretical Dimensions and Properties for **Designing**

Section Number	Weight per Foot	Area of Section A	Depth of Section d	Flange Width b_f	Flange Thickness t_f	Stem Thickness t_w	Axis X-X I_x	Axis X-X S_x	Axis X-X r_x	Axis X-X y	Axis Y-Y I_y	Axis Y-Y S_y	Axis Y-Y r_y
	lb	in.²	in.	in.	in.	in.	in.⁴	in.³	in.	in.	in.⁴	in.³	in.
WT7 x	66	19.4	7.330	14.725	1.030	0.645	57.8	9.57	1.73	1.29	274	37.2	3.76
	60	17.7	7.240	14.670	0.940	0.590	51.7	8.61	1.71	1.24	247	33.7	3.74
	54.5	16.0	7.160	14.605	0.860	0.525	45.3	7.56	1.68	1.17	223	30.6	3.73
	49.5	14.6	7.080	14.565	0.780	0.485	40.9	6.88	1.67	1.14	201	27.6	3.71
	45	13.2	7.010	14.520	0.710	0.440	36.4	6.16	1.66	1.09	181	25.0	3.70
WT7 x	41	12.0	7.155	10.130	0.855	0.510	41.2	7.14	1.85	1.39	74.2	14.6	2.48
	37	10.9	7.085	10.070	0.785	0.450	36.0	6.25	1.82	1.32	66.9	13.3	2.48
	34	9.99	7.020	10.035	0.720	0.415	32.6	5.69	1.81	1.29	60.7	12.1	2.46
	30.5	8.96	6.945	9.995	0.645	0.375	28.9	5.07	1.80	1.25	53.7	10.7	2.45
WT7 x	26.5	7.81	6.960	8.060	0.660	0.370	27.6	4.94	1.88	1.38	28.8	7.16	1.92
	24	7.07	6.895	8.030	0.595	0.340	24.9	4.48	1.87	1.35	25.7	6.40	1.91
	21.5	6.31	6.830	7.995	0.530	0.305	21.9	3.98	1.86	1.31	22.6	5.65	1.89
WT7 x	19	5.58	7.050	6.770	0.515	0.310	23.3	4.22	2.04	1.54	13.3	3.94	1.55
	17	5.00	6.990	6.745	0.455	0.285	20.9	3.83	2.04	1.53	11.7	3.45	1.53
	15	4.42	6.920	6.730	0.385	0.270	19.0	3.55	2.07	1.58	9.79	2.91	1.49
WT7 x	13	3.85	6.955	5.025	0.420	0.255	17.3	3.31	2.12	1.72	4.45	1.77	1.08
	11	3.25	6.870	5.000	0.335	0.230	14.8	2.91	2.14	1.76	3.50	1.40	1.04
WT6 x	95	27.9	7.190	12.670	1.735	1.060	79.0	14.2	1.68	1.62	295	46.5	3.25
	85	25.0	7.015	12.570	1.560	0.960	67.8	12.3	1.65	1.52	259	41.2	3.22
	76	22.4	6.855	12.480	1.400	0.870	58.5	10.8	1.62	1.43	227	36.4	3.19
	68	20.0	6.705	12.400	1.250	0.790	50.6	9.46	1.59	1.35	199	32.1	3.16
	60	17.6	6.560	12.320	1.105	0.710	43.4	8.22	1.57	1.28	172	28.0	3.13
	53	15.6	6.445	12.220	0.990	0.610	36.3	6.91	1.53	1.19	151	24.7	3.11
	48	14.1	6.355	12.160	0.900	0.550	32.0	6.12	1.51	1.13	135	22.2	3.09
	43.5	12.8	6.265	12.125	0.810	0.515	28.9	5.60	1.50	1.10	120	19.9	3.07
	39.5	11.6	6.190	12.080	0.735	0.470	25.8	5.03	1.49	1.06	108	17.9	3.05
	36	10.6	6.125	12.040	0.670	0.430	23.2	4.54	1.48	1.02	97.5	16.2	3.04
	32.5	9.54	6.060	12.000	0.605	0.390	20.6	4.06	1.47	0.985	87.2	14.5	3.02

Properties shown in this table are for the full center split section.

Taken from Bethlehem Steel *Structural Shapes* handbook.

STRUCTURAL TEES (Cut from W Shapes)

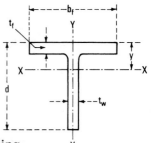

Theoretical Dimensions and Properties for Designing

Section Number	Weight per Foot (lb)	Area of Section A (in.²)	Depth of Section d (in.)	Flange Width b_f (in.)	Flange Thickness t_f (in.)	Stem Thickness t_w (in.)	Axis X-X I_x (in.⁴)	Axis X-X S_x (in.³)	Axis X-X r_x (in.)	Axis X-X y (in.)	Axis Y-Y I_y (in.⁴)	Axis Y-Y S_y (in.³)	Axis Y-Y r_y (in.)
WT6 x	29	8.52	6.095	10.010	0.640	0.360	19.1	3.76	1.50	1.03	53.5	10.7	2.51
	26.5	7.78	6.030	9.995	0.575	0.345	17.7	3.54	1.51	1.02	47.9	9.58	2.48
WT6 x	25	7.34	6.095	8.080	0.640	0.370	18.7	3.79	1.60	1.17	28.2	6.97	1.96
	22.5	6.61	6.030	8.045	0.575	0.335	16.6	3.39	1.58	1.13	25.0	6.21	1.94
	20	5.89	5.970	8.005	0.515	0.295	14.4	2.95	1.57	1.08	22.0	5.51	1.93
WT6 x	17.5	5.17	6.250	6.560	0.520	0.300	16.0	3.23	1.76	1.30	12.2	3.73	1.54
	15	4.40	6.170	6.520	0.440	0.260	13.5	2.75	1.75	1.27	10.2	3.12	1.52
	13	3.82	6.110	6.490	0.380	0.230	11.7	2.40	1.75	1.25	8.66	2.67	1.51
WT6 x	11	3.24	6.155	4.030	0.425	0.260	11.7	2.59	1.90	1.63	2.33	1.16	0.848
	9.5	2.79	6.080	4.005	0.350	0.235	10.1	2.28	1.90	1.65	1.88	0.939	0.822
	8	2.36	5.995	3.990	0.265	0.220	8.70	2.04	1.92	1.74	1.41	0.706	0.773
	7	2.08	5.955	3.970	0.225	0.200	7.67	1.83	1.92	1.76	1.18	0.594	0.753
WT5 x	56	16.5	5.680	10.415	1.250	0.755	28.6	6.40	1.32	1.21	118	22.6	2.68
	50	14.7	5.550	10.340	1.120	0.680	24.5	5.56	1.29	1.13	103	20.0	2.65
	44	12.9	5.420	10.265	0.990	0.605	20.8	4.77	1.27	1.06	89.3	17.4	2.63
	38.5	11.3	5.300	10.190	0.870	0.530	17.4	4.05	1.24	0.990	76.8	15.1	2.60
	34	9.99	5.200	10.130	0.770	0.470	14.9	3.49	1.22	0.932	66.8	13.2	2.59
	30	8.82	5.110	10.080	0.680	0.420	12.9	3.04	1.21	0.884	58.1	11.5	2.57
	27	7.91	5.045	10.030	0.615	0.370	11.1	2.64	1.19	0.836	51.7	10.3	2.56
	24.5	7.21	4.990	10.000	0.560	0.340	10.0	2.39	1.18	0.807	46.7	9.34	2.54
WT5 x	22.5	6.63	5.050	8.020	0.620	0.350	10.2	2.47	1.24	0.907	26.7	6.65	2.01
	19.5	5.73	4.960	7.985	0.530	0.315	8.84	2.16	1.24	0.876	22.5	5.64	1.98
	16.5	4.85	4.865	7.960	0.435	0.290	7.71	1.93	1.26	0.869	18.3	4.60	1.94
WT5 x	15	4.42	5.235	5.810	0.510	0.300	9.28	2.24	1.45	1.10	8.35	2.87	1.37
	13	3.81	5.165	5.770	0.440	0.260	7.86	1.91	1.44	1.06	7.05	2.44	1.36
	11	3.24	5.085	5.750	0.360	0.240	6.88	1.72	1.46	1.07	5.71	1.99	1.33
WT5 x	9.5	2.81	5.120	4.020	0.395	0.250	6.68	1.74	1.54	1.28	2.15	1.07	0.874
	8.5	2.50	5.055	4.010	0.330	0.240	6.06	1.62	1.56	1.32	1.78	0.888	0.845
	7.5	2.21	4.995	4.000	0.270	0.230	5.45	1.50	1.57	1.37	1.45	0.723	0.810
	6	1.77	4.935	3.960	0.210	0.190	4.35	1.22	1.57	1.36	1.09	0.551	0.785

Properties shown in this table are for the full center split section.

Taken from Bethlehem Steel *Structural Shapes* handbook.

STRUCTURAL TEES (Cut from W Shapes)

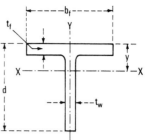

Theoretical Dimensions and Properties for **Designing**

Section Number	Weight per Foot	Area of Section	Depth of Section	Flange Width	Flange Thickness	Stem Thickness	Axis X-X I_x	Axis X-X S_x	Axis X-X r_x	Axis X-X y	Axis Y-Y I_y	Axis Y-Y S_y	Axis Y-Y r_y
		A	d	b_f	t_f	t_w							
	lb	in.²	in.	in.	in.	in.	in.⁴	in.³	in.	in.	in.⁴	in.³	in.
WT4 x	33.5	9.84	4.500	8.280	0.935	0.570	10.9	3.05	1.05	0.936	44.3	10.7	2.12
	29	8.55	4.375	8.220	0.810	0.510	9.12	2.61	1.03	0.874	37.5	9.13	2.10
	24	7.05	4.250	8.110	0.685	0.400	6.85	1.97	0.986	0.777	30.5	7.52	2.08
	20	5.87	4.125	8.070	0.560	0.360	5.73	1.69	0.988	0.735	24.5	6.08	2.04
	17.5	5.14	4.060	8.020	0.495	0.310	4.81	1.43	0.968	0.688	21.3	5.31	2.03
	15.5	4.56	4.000	7.995	0.435	0.285	4.28	1.28	0.968	0.668	18.5	4.64	2.02
WT4 x	14	4.12	4.030	6.535	0.465	0.285	4.22	1.28	1.01	0.734	10.8	3.31	1.62
	12	3.54	3.965	6.495	0.400	0.245	3.53	1.08	0.999	0.695	9.14	2.81	1.61
WT4 x	10.5	3.08	4.140	5.270	0.400	0.250	3.90	1.18	1.12	0.831	4.89	1.85	1.26
	9	2.63	4.070	5.250	0.330	0.230	3.41	1.05	1.14	0.834	3.98	1.52	1.23
WT4 x	7.5	2.22	4.055	4.015	0.315	0.245	3.28	1.07	1.22	0.998	1.70	0.849	0.876
	6.5	1.92	3.995	4.000	0.255	0.230	2.89	0.974	1.23	1.03	1.37	0.683	0.844
	5	1.48	3.945	3.940	0.205	0.170	2.15	0.717	1.20	0.953	1.05	0.532	0.841
WT3 x	12.5	3.67	3.190	6.080	0.455	0.320	2.29	0.886	0.789	0.610	8.53	2.81	1.52
	10	2.94	3.100	6.020	0.365	0.260	1.76	0.693	0.774	0.560	6.64	2.21	1.50
	7.5	2.21	2.995	5.990	0.260	0.230	1.41	0.577	0.797	0.558	4.66	1.56	1.45
WT3 x	8	2.37	3.140	4.030	0.405	0.260	1.69	0.685	0.844	0.677	2.21	1.10	0.967
	6	1.78	3.015	4.000	0.280	0.230	1.32	0.564	0.862	0.677	1.50	0.748	0.918
	4.5	1.34	2.950	3.940	0.215	0.170	0.950	0.408	0.842	0.623	1.10	0.557	0.905
WT2.5 x	9.5	2.77	2.575	5.030	0.430	0.270	1.01	0.485	0.605	0.487	4.56	1.82	1.28
	8	2.34	2.505	5.000	0.360	0.240	0.845	0.413	0.601	0.458	3.75	1.50	1.27
WT2 x	6.5	1.91	2.080	4.060	0.345	0.280	0.526	0.321	0.524	0.440	1.93	0.950	1.00

Properties shown in this table are for the full center split section.

Taken from Bethlehem Steel *Structural Shapes* handbook.

STRUCTURAL TEE (Cut from M Shape)

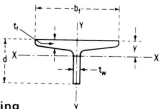

Theoretical Dimensions and Properties for **Designing**

Section Number	Weight per Foot	Area of Section A	Depth of Section d	Flange Width b_f	Flange Thickness t_f	Stem Thickness t_w	Axis X-X I_x	Axis X-X S_x	Axis X-X r_x	Axis X-X y	Axis Y-Y I_y	Axis Y-Y S_y	Axis Y-Y r_y
	lb	in.²	in.	in.	in.	in.	in.⁴	in.³	in.	in.⁴	in.³	in.	in.
MT2.5 x	9.45	2.78	2.500	5.003	0.416*	0.316	1.05	0.527	0.615	0.511	3.93	1.57	1.19

*Average thickness

Taken from Bethlehem Steel *Structural Shapes* handbook.

AMERICAN STANDARD SHAPES

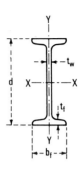

Theoretical Dimensions and Properties for **Designing**

| Section Number | Weight per Foot | Area of Section A | Depth of Section d | Flange | | Web Thickness t_w | Axis X-X | | | Axis Y-Y | | | r_T |
| | | | | Width b_f | Average Thickness t_f | | I_x | S_x | r_x | I_y | S_y | r_y | |
	lb	in.²	in.	in.	in.	in.	in.⁴	in.³	in.	in.⁴	in.³	in.	in.
S24 x	121.0	35.6	24.50	8.050	1.090	0.800	3160	258	9.43	83.3	20.7	1.53	1.86
	106.0	31.2	24.50	7.870	1.090	0.620	2940	240	9.71	77.1	19.6	1.57	1.86
S24 x	100.0	29.3	24.00	7.245	0.870	0.745	2390	199	9.02	47.7	13.2	1.27	1.59
	90.0	26.5	24.00	7.125	0.870	0.625	2250	187	9.21	44.9	12.6	1.30	1.60
	80.0	23.5	24.00	7.000	0.870	0.500	2100	175	9.47	42.2	12.1	1.34	1.61
S20 x	96.0	28.2	20.30	7.200	0.920	0.800	1670	165	7.71	50.2	13.9	1.33	1.63
	86.0	25.3	20.30	7.060	0.920	0.660	1580	155	7.89	46.8	13.3	1.36	1.63
S20 x	75.0	22.0	20.00	6.385	0.795	0.635	1280	128	7.62	29.8	9.32	1.16	1.43
	66.0	19.4	20.00	6.255	0.795	0.505	1190	119	7.83	27.7	8.85	1.19	1.44
S18 x	70.0	20.6	18.00	6.251	0.691	0.711	926	103	6.71	24.1	7.72	1.08	1.40
	54.7	16.1	18.00	6.001	0.691	0.461	804	89.4	7.07	20.8	6.94	1.14	1.40
S15 x	50.0	14.7	15.00	5.640	0.622	0.550	486	64.8	5.75	15.7	5.57	1.03	1.30
	42.9	12.6	15.00	5.501	0.622	0.411	447	59.6	5.95	14.4	5.23	1.07	1.30
S12 x	50.0	14.7	12.00	5.477	0.659	0.687	305	50.8	4.55	15.7	5.74	1.03	1.31
	40.8	12.0	12.00	5.252	0.659	0.462	272	45.4	4.77	13.6	5.16	1.06	1.28
S12 x	35.0	10.3	12.00	5.078	0.544	0.428	229	38.2	4.72	9.87	3.89	0.980	1.20
	31.8	9.35	12.00	5.000	0.544	0.350	218	36.4	4.83	9.36	3.74	1.00	1.20

All shapes on these pages have a flange slope of 16⅔ pct.

Taken from Bethlehem Steel *Structural Shapes* handbook.

AMERICAN STANDARD SHAPES

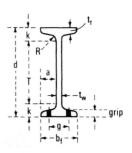

Approximate Dimensions for **Detailing**

Section Number	Weight per Foot	Depth of Section	Flange		Web Thickness	Half Web Thickness	a	T	k	R	Grip	Max Flange Fastener	Usual Flange Gage
			Width	Average Thickness									
		d	b_f	t_f	t_w	$\frac{t_w}{2}$							g
	lb	in.	in.	in.	in.	in.	in.	in.	in.	in.	in.	in.	in.
S24 x	121.0	24½	8	1 1/16	13/16	3/8	3 5/8	20½	2	.60	1 1/8	1	4
	106.0	24½	7 7/8	1 1/16	5/8	5/16	3 5/8	20½	2	.60	1 1/8	1	4
S24 x	100.0	24	7¼	7/8	3/4	3/8	3¼	20½	1¾	.60	7/8	1	4
	90.0	24	7 1/8	7/8	5/8	5/16	3¼	20½	1¾	.60	7/8	1	4
	80.0	24	7	7/8	1/2	1/4	3¼	20½	1¾	.60	7/8	1	4
S20 x	96.0	20¼	7¼	15/16	13/16	3/8	3¼	16¾	1¾	.60	15/16	1	4
	86.0	20¼	7	15/16	11/16	5/16	3¼	16¾	1¾	.60	15/16	1	4
S20 x	75.0	20	6 3/8	13/16	5/8	5/16	2 7/8	16¾	1 5/8	.60	13/16	7/8	3½
	66.0	20	6¼	13/16	1/2	1/4	2 7/8	16¾	1 5/8	.60	13/16	7/8	3½
S18 x	70.0	18	6¼	11/16	11/16	3/8	2¾	15	1½	.56	11/16	7/8	3½
	54.7	18	6	11/16	7/16	1/4	2¾	15	1½	.56	11/16	7/8	3½
S15 x	50.0	15	5 5/8	5/8	9/16	1/4	2½	12¼	1 3/8	.51	9/16	3/4	3½
	42.9	15	5½	5/8	7/16	3/16	2½	12¼	1 3/8	.51	9/16	3/4	3½
S12 x	50.0	12	5½	11/16	11/16	5/16	2 3/8	9 1/8	1 7/16	.56	11/16	3/4	3
	40.8	12	5¼	11/16	7/16	1/4	2 3/8	9 1/8	1 7/16	.56	5/8	3/4	3
S12 x	35.0	12	5 1/8	9/16	7/16	3/16	2 3/8	9 5/8	1 3/16	.45	1/2	3/4	3
	31.8	12	5	9/16	3/8	3/16	2 3/8	9 5/8	1 3/16	.45	1/2	3/4	3

Taken from Bethlehem Steel *Structural Shapes* handbook.

AMERICAN STANDARD SHAPES

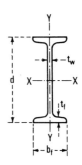

Theoretical Dimensions and Properties for **Designing**

Section Number	Weight per Foot	Area of Section A	Depth of Section d	Flange		Web Thickness t_w	Axis X-X			Axis Y-Y			r_T
				Width b_f	Average Thickness t_f		I_x	S_x	r_x	I_y	S_y	r_y	
	lb	in.²	in.	in.	in.	in.	in.⁴	in.³	in.	in.⁴	in.³	in.	in.
S10 x	35.0	10.3	10.00	4.944	0.491	0.594	147	29.4	3.78	8.36	3.38	0.901	1.14
	25.4	7.46	10.00	4.661	0.491	0.311	124	24.7	4.07	6.79	2.91	0.954	1.12
S8 x	23.0	6.77	8.00	4.171	0.425	0.441	64.9	16.2	3.10	4.31	2.07	0.798	0.984
	18.4	5.41	8.00	4.001	0.425	0.271	57.6	14.4	3.26	3.73	1.86	0.831	0.969
S7 x	15.3	4.50	7.00	3.662	0.392	0.252	36.7	10.5	2.86	2.64	1.44	0.766	0.891
S6 x	17.25	5.07	6.00	3.565	0.359	0.465	26.3	8.77	2.28	2.31	1.30	0.675	0.842
	12.5	3.67	6.00	3.332	0.359	0.232	22.1	7.37	2.45	1.82	1.09	0.705	0.814

All shapes on these pages have a flange slope of 16⅔ pct.

Taken from Bethlehem Steel *Structural Shapes* handbook.

AMERICAN STANDARD SHAPES

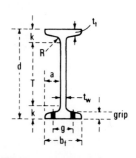

Approximate Dimensions for **Detailing**

Section Number	Weight per Foot	Depth of Section	Flange		Web Thickness	Half Web Thickness	a	T	k	R	Grip	Max Flange Fastener	Usual Flange Gage
			Width	Average Thickness									
		d	b_f	t_f	t_w	$\frac{t_w}{2}$							g
	lb	in.	in.	in.	in.	in.	in.	in.	in.	in.	in.	in.	in.
S10 x	35.0	10	5	½	⅝	5/16	2⅛	7¾	1⅛	.41	½	¾	2¾
	25.4	10	4⅝	½	5/16	⅛	2⅛	7¾	1⅛	.41	½	¾	2¾
S8 x	23.0	8	4⅛	7/16	7/16	¼	1⅞	6	1	.37	7/16	¾	2¼
	18.4	8	4	7/16	¼	⅛	1⅞	6	1	.37	7/16	¾	2¼
S7 x	15.3	7	3⅝	⅜	¼	⅛	1¾	5¼	⅞	.35	⅜	⅝	2¼
S6 x	17.25	6	3⅝	⅜	7/16	¼	1½	4⅜	13/16	.33	⅜	⅝	2
	12.5	6	3⅜	⅜	¼	⅛	1½	4⅜	13/16	.33	5/16	—	—

Taken from Bethlehem Steel *Structural Shapes* handbook.

AMERICAN STANDARD CHANNELS

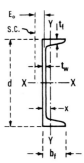

Theoretical Dimensions and Properties for **Designing**

Section Number	Weight per Foot	Area of Section	Depth of Section	Flange		Web Thickness	Axis X-X			Axis Y-Y				Shear Center Location
				Width	Average Thickness									
		A	d	b_f	t_f	t_w	I_x	S_x	r_x	I_y	S_y	r_y	x	E_o
	lb	in.²	in.	in.	in.	in.	in.⁴	in.³	in.	in.⁴	in.³	in.	in.	in.
C15 x	50.0	14.7	15.00	3.716	0.650	0.716	404	53.8	5.24	11.0	3.78	0.867	0.799	0.941
	40.0	11.8	15.00	3.520	0.650	0.520	349	46.5	5.44	9.23	3.36	0.886	0.778	1.03
	33.9	9.96	15.00	3.400	0.650	0.400	315	42.0	5.62	8.13	3.11	0.904	0.787	1.10
C12 x	30.0	8.82	12.00	3.170	0.501	0.510	162	27.0	4.29	5.14	2.06	0.763	0.674	0.873
	25.0	7.35	12.00	3.047	0.501	0.387	144	24.1	4.43	4.47	1.88	0.780	0.674	0.940
	20.7	6.09	12.00	2.942	0.501	0.282	129	21.5	4.61	3.88	1.73	0.799	0.698	1.01
C10 x	30.0	8.82	10.00	3.033	0.436	0.673	103	20.7	3.42	3.94	1.65	0.669	0.649	0.705
	25.0	7.35	10.00	2.886	0.436	0.526	91.2	18.2	3.52	3.36	1.48	0.676	0.617	0.757
	20.0	5.88	10.00	2.739	0.436	0.379	78.9	15.8	3.66	2.81	1.32	0.691	0.606	0.826
	15.3	4.49	10.00	2.600	0.436	0.240	67.4	13.5	3.87	2.28	1.16	0.713	0.634	0.916
C9 x	15.0	4.41	9.00	2.485	0.413	0.285	51.0	11.3	3.40	1.93	1.01	0.661	0.586	0.824
	13.4	3.94	9.00	2.433	0.413	0.233	47.9	10.6	3.48	1.76	0.962	0.668	0.601	0.859
C8 x	18.75	5.51	8.00	2.527	0.390	0.487	44.0	11.0	2.82	1.98	1.01	0.599	0.565	0.674
	13.75	4.04	8.00	2.343	0.390	0.303	36.1	9.03	2.99	1.53	0.853	0.615	0.553	0.756
	11.5	3.38	8.00	2.260	0.390	0.220	32.6	8.14	3.11	1.32	0.781	0.625	0.571	0.807
C7 x	12.25	3.60	7.00	2.194	0.366	0.314	24.2	6.93	2.60	1.17	0.702	0.571	0.525	0.695
	9.8	2.87	7.00	2.090	0.366	0.210	21.3	6.08	2.72	0.968	0.625	0.581	0.541	0.752

All shapes on these pages have a flange slope of 16²/₃ pct.

Taken from Bethlehem Steel *Structural Shapes* handbook.

AMERICAN STANDARD CHANNELS

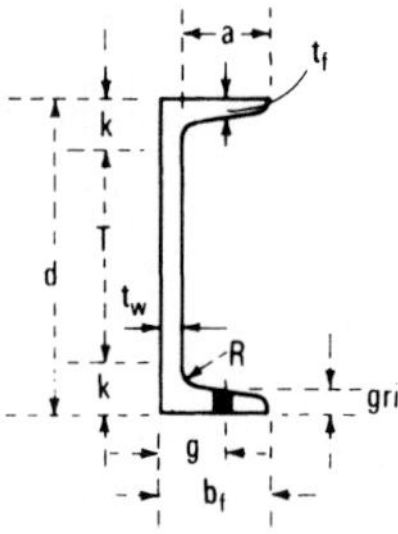

Approximate Dimensions for **Detailing**

Section Number	Weight per Foot	Depth of Section	Flange: Width	Flange: Average Thickness	Web Thickness	Half Web Thickness	a	T	k	R	Grip	Max Flange Fastener	Usual Flange Gage
		d	b_f	t_f	t_w	$\frac{t_w}{2}$							g
	lb	in.	in.	in.	in.	in.	in.	in.	in.	in.	in.	in.	in.
C15 x	**50.0**	15	3 3/4	5/8	11/16	3/8	3	12 1/8	1 7/16	0.50	5/8	1	2 1/4
	40.0	15	3 1/2	5/8	1/2	1/4	3	12 1/8	1 7/16	0.50	5/8	1	2
	33.9	15	3 3/8	5/8	3/8	3/16	3	12 1/8	1 7/16	0.50	5/8	1	2
C12 x	**30.0**	12	3 1/8	1/2	1/2	1/4	2 5/8	9 3/4	1 1/8	0.38	1/2	7/8	1 3/4
	25.0	12	3	1/2	3/8	3/16	2 5/8	9 3/4	1 1/8	0.38	1/2	7/8	1 3/4
	20.7	12	3	1/2	5/16	1/8	2 5/8	9 3/4	1 1/8	0.38	1/2	7/8	1 3/4
C10 x	**30.0**	10	3	7/16	11/16	5/16	2 3/8	8	1	0.34	7/16	3/4	1 3/4
	25.0	10	2 7/8	7/16	1/2	1/4	2 3/8	8	1	0.34	7/16	3/4	1 3/4
	20.0	10	2 3/4	7/16	3/8	3/16	2 3/8	8	1	0.34	7/16	3/4	1 1/2
	15.3	10	2 5/8	7/16	1/4	1/8	2 3/8	8	1	0.34	7/16	3/4	1 1/2
C9 x	**15.0**	9	2 1/2	7/16	5/16	1/8	2 1/4	7 1/8	15/16	0.33	7/16	3/4	1 3/8
	13.4	9	2 3/8	7/16	1/4	1/8	2 1/4	7 1/8	15/16	0.33	7/16	3/4	1 3/8
C8 x	**18.75**	8	2 1/2	3/8	1/2	1/4	2	6 1/8	15/16	0.32	3/8	3/4	1 1/2
	13.75	8	2 3/8	3/8	5/16	1/8	2	6 1/8	15/16	0.32	3/8	3/4	1 3/8
	11.5	8	2 1/4	3/8	1/4	1/8	2	6 1/8	15/16	0.32	3/8	3/4	1 3/8
C7 x	**12.25**	7	2 1/4	3/8	5/16	3/16	1 7/8	5 1/4	7/8	0.31	3/8	5/8	1 1/4
	9.8	7	2 1/8	3/8	3/16	1/8	1 7/8	5 1/4	7/8	0.31	3/8	5/8	1 1/4

Taken from Bethlehem Steel *Structural Shapes* handbook.

MISCELLANEOUS CHANNELS[W]

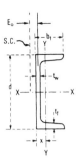

Theoretical Dimensions and Properties for **Designing**

Section Number	Weight per Foot	Area of Section A	Depth of Section d	Flange Width b_f	Flange Average Thickness t_f	Web Thickness t_w	Axis X-X I_x	Axis X-X S_x	Axis X-X r_x	Axis Y-Y I_y	Axis Y-Y S_y	Axis Y-Y r_y	Axis Y-Y x	Shear Center Location E_o
	lb	in.²	in.	in.	in.	in.	in.⁴	in.³	in.	in.⁴	in.³	in.	in.	in.
[x]MC18 x	58.0	17.1	18.00	4.200	0.625	0.700	676	75.1	6.29	17.8	5.32	1.02	0.862	1.04
	51.9	15.3	18.00	4.100	0.625	0.600	627	69.7	6.41	16.4	5.07	1.04	0.858	1.10
	45.8	13.5	18.00	4.000	0.625	0.500	578	64.3	6.56	15.1	4.82	1.06	0.866	1.16
	42.7	12.6	18.00	3.950	0.625	0.450	554	61.6	6.64	14.4	4.69	1.07	0.877	1.19
[y]MC13 x	50.0	14.7	13.00	4.412	0.610	0.787	314	48.4	4.62	16.5	4.79	1.06	0.974	1.21
	40.0	11.8	13.00	4.185	0.610	0.560	273	42.0	4.82	13.7	4.26	1.08	0.964	1.31
	35.0	10.3	13.00	4.072	0.610	0.447	252	38.8	4.95	12.3	3.99	1.10	0.980	1.38
	31.8	9.35	13.00	4.000	0.610	0.375	239	36.8	5.06	11.4	3.81	1.11	1.00	1.43
[z]MC12 x	50.0	14.7	12.00	4.135	0.700	0.835	269	44.9	4.28	17.4	5.65	1.09	1.05	1.16
	45.0	13.2	12.00	4.012	0.700	0.712	252	42.0	4.36	15.8	5.33	1.09	1.04	1.20
	40.0	11.8	12.00	3.890	0.700	0.590	234	39.0	4.46	14.3	5.00	1.10	1.04	1.25
	35.0	10.3	12.00	3.767	0.700	0.467	216	36.1	4.59	12.7	4.67	1.11	1.05	1.30
	31.0	9.12	12.00	3.670	0.700	0.370	203	33.8	4.71	11.3	4.39	1.12	1.08	1.36
[x]MC10 x	41.1	12.1	10.00	4.321	0.575	0.796	158	31.5	3.61	15.8	4.88	1.14	1.09	1.26
	33.6	9.87	10.00	4.100	0.575	0.575	139	27.8	3.75	13.2	4.38	1.16	1.08	1.35
	28.5	8.37	10.00	3.950	0.575	0.425	127	25.3	3.89	11.4	4.02	1.17	1.12	1.43
[x]MC10 x	25.0	7.35	10.00	3.405	0.575	0.380	110	22.0	3.87	7.35	3.00	1.00	0.953	1.22
	22.0	6.45	10.00	3.315	0.575	0.290	103	20.5	3.99	6.50	2.80	1.00	0.990	1.27
[x]MC9 x	25.4	7.47	9.00	3.500	0.550	0.450	88.0	19.6	3.43	7.65	3.02	1.01	0.970	1.21
	23.9	7.02	9.00	3.450	0.550	0.400	85.0	18.9	3.48	7.22	2.93	1.01	0.981	1.24

[W] All sections on these pages are special sections. See page 3.
[x] These channels have a flange slope of 2 degrees.
[y] These channels have a flange slope of 8.5 degrees.
[z] These channels have a flange slope of 1.7 degrees.

Taken from Bethlehem Steel *Structural Shapes* handbook.

MISCELLANEOUS CHANNELS[w]

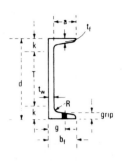

Approximate Dimensions for **Detailing**

Section Number	Weight per Foot	Depth of Section	Flange Width	Flange Average Thickness	Web Thickness	Half Web Thickness	a	T	k	R	Grip	Max Flange Fastener	Usual Flange Gage
		d	b_f	t_f	t_w	$\frac{t_w}{2}$							g
	lb	in.	in.	in.	in.	in.	in.	in.	in.	in.	in.	in.	in.
[x]MC18 x	58.0	18	4 1/4	5/8	11/16	3/8	3 1/2	15 1/4	1 3/8	0.625	5/8	1	2 1/2
	51.9	18	4 1/8	5/8	5/8	5/16	3 1/2	15 1/4	1 3/8	0.625	5/8	1	2 1/2
	45.8	18	4	5/8	1/2	1/4	3 1/2	15 1/4	1 3/8	0.625	5/8	1	2 1/2
	42.7	18	4	5/8	7/16	1/4	3 1/2	15 1/4	1 3/8	0.625	5/8	1	2 1/2
[y]MC13 x	50.0	13	4 3/8	5/8	13/16	3/8	3 5/8	10 1/4	1 3/8	0.48	5/8	1	2 1/2
	40.0	13	4 1/8	5/8	9/16	1/4	3 5/8	10 1/4	1 3/8	0.48	9/16	1	2 1/2
	35.0	13	4 1/8	5/8	7/16	1/4	3 5/8	10 1/4	1 3/8	0.48	9/16	1	2 1/2
	31.8	13	4	5/8	3/8	3/16	3 5/8	10 1/4	1 3/8	0.48	9/16	1	2 1/2
[z]MC12 x	50.0	12	4 1/8	11/16	13/16	7/16	3 1/4	9 3/8	1 5/16	0.50	11/16	1	2 1/2
	45.0	12	4	11/16	11/16	3/8	3 1/4	9 3/8	1 5/16	0.50	11/16	1	2 1/2
	40.0	12	3 7/8	11/16	9/16	5/16	3 1/4	9 3/8	1 5/16	0.50	11/16	1	2 1/2
	35.0	12	3 3/4	11/16	7/16	1/4	3 1/4	9 3/8	1 5/16	0.50	11/16	1	2 1/2
	31.0	12	3 5/8	11/16	3/8	3/16	3 1/4	9 3/8	1 5/16	0.50	11/16	1	2 1/2
[x]MC10 x	41.1	10	4 3/8	9/16	13/16	3/8	3 1/2	7 1/2	1 1/4	0.575	9/16	7/8	2 1/2
	33.6	10	4 1/8	9/16	9/16	5/16	3 1/2	7 1/2	1 1/4	0.575	9/16	7/8	2 1/2
	28.5	10	4	9/16	7/16	3/16	3 1/2	7 1/2	1 1/4	0.575	9/16	7/8	2 1/2
[x]MC10 x	25.0	10	3 3/8	9/16	3/8	3/16	3	7 1/2	1 1/4	0.570	9/16	7/8	2
	22.0	10	3 3/8	9/16	5/16	1/8	3	7 1/2	1 1/4	0.570	9/16	7/8	2
[x]MC9 x	25.4	9	3 1/2	9/16	7/16	1/4	3	6 5/8	1 3/16	0.550	9/16	7/8	2
	23.9	9	3 1/2	9/16	3/8	3/16	3	6 5/8	1 3/16	0.550	9/16	7/8	2

Taken from Bethlehem Steel *Structural Shapes* handbook.

MISCELLANEOUS CHANNELS[W]

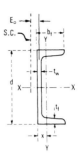

Theoretical Dimensions and Properties for **Designing**

Section Number		Weight per Foot	Area of Section	Depth of Section	Flange		Web Thickness	Axis X-X			Axis Y-Y				Shear Center Location E_o
					Width b_f	Average Thickness t_f	t_w	I_x	S_x	r_x	I_y	S_y	r_y	x	
		lb	in.²	in.	in.	in.	in.	in.⁴	in.³	in.	in.⁴	in.³	in.	in.	in.
˄MC8	x	22.8	6.70	8.00	3.502	0.525	0.427	63.8	16.0	3.09	7.07	2.84	1.03	1.01	1.26
		21.4	6.28	8.00	3.450	0.525	0.375	61.6	15.4	3.13	6.64	2.74	1.03	1.02	1.28
˄MC8	x	20.0	5.88	8.00	3.025	0.500	0.400	54.5	13.6	3.05	4.47	2.05	0.872	0.840	1.04
		18.7	5.50	8.00	2.978	0.500	0.353	52.5	13.1	3.09	4.20	1.97	0.874	0.849	1.07
˄MC7	x	ʸ22.7	6.67	7.00	3.603	0.500	0.503	47.5	13.6	2.67	7.29	2.85	1.05	1.04	1.26
		ʸ19.1	5.61	7.00	3.452	0.500	0.352	43.2	12.3	2.77	6.11	2.57	1.04	1.08	1.33
˄MC6	x	18.0	5.29	6.00	3.504	0.475	0.379	29.7	9.91	2.37	5.93	2.48	1.06	1.12	1.36
˄MC6	x	15.3	4.50	6.00	3.500	0.385	0.340	25.4	8.47	2.38	4.97	2.03	1.05	1.05	1.33
˄MC6	x	16.3	4.79	6.00	3.000	0.475	0.375	26.0	8.68	2.33	3.82	1.84	0.892	0.927	1.12
		15.1	4.44	6.00	2.941	0.475	0.316	25.0	8.32	2.37	3.51	1.75	0.889	0.940	1.14
˄MC6	x	12.0	3.53	6.00	2.497	0.375	0.310	18.7	6.24	2.30	1.87	1.04	0.728	0.704	0.880

˄All sections on these pages are special sections. See page 3.
˄Available subject to inquiry.
ʸThese channels have a flange slope of 2 degrees.

Taken from Bethlehem Steel *Structural Shapes* handbook.

MISCELLANEOUS CHANNELS[W]

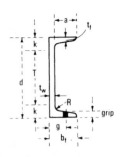

Approximate Dimensions for **Detailing**

Section Number	Weight per Foot	Depth of Section	Flange Width b_f	Flange Average Thickness t_f	Web Thickness t_w	Half Web Thickness $\frac{t_w}{2}$	a	T	k	R	Grip	Max Flange Fastener	Usual Flange Gage g
	lb	in.	in.	in.	in.	in.	in.	in.	in.	in.	in.	in.	in.
MC8 x 22.8		8	3½	½	7/16	3/16	3⅛	5⅝	1 3/16	0.525	½	⅞	2
21.4		8	3½	½	⅜	3/16	3⅛	5⅝	1 3/16	0.525	½	⅞	2
MC8 x 20.0		8	3	½	⅜	3/16	2⅝	5¾	1⅛	0.500	½	⅞	2
18.7		8	3	½	⅜	3/16	2⅝	5¾	1⅛	0.500	½	⅞	2
MC7 x 22.7		7	3⅝	½	½	¼	3⅛	4¾	1⅛	0.500	½	⅞	2
19.1		7	3½	½	⅜	3/16	3⅛	4¾	1⅛	0.500	½	⅞	2
MC6 x 18.0		6	3½	½	⅜	3/16	3⅛	3⅞	1 1/16	0.475	½	⅞	2
MC6 x 15.3		6	3½	⅜	5/16	3/16	3⅛	4¼	⅞	0.385	⅜	⅞	2
MC6 x 16.3		6	3	½	⅜	3/16	2⅝	3⅞	1 1/16	0.475	½	¾	1¾
15.1		6	3	½	5/16	3/16	2⅝	3⅞	1 1/16	0.475	½	¾	1¾
MC6 x 12.0		6	2½	⅜	5/16	⅛	2⅛	4⅜	13/16	0.375	⅜	⅝	1½

Taken from Bethlehem Steel *Structural Shapes* handbook.

STRUCTURAL TEES (Cut from S Shapes)

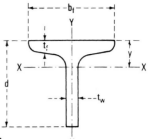

Theoretical Dimensions and Properties for **Designing**

Section Number	Weight per Foot	Area of Section	Depth of Tee	Flange		Stem Thickness	Axis X-X				Axis Y-Y		
				Width	Average Thickness								
		A	d	b_f	t_f	t_w	I_x	S_x	r_x	y	I_y	S_y	r_y
	lb	in.²	in.	in.	in.	in.	in.⁴	in.³	in.	in.	in.⁴	in.³	in.
ST12 x	60.5	17.8	12.250	8.050	1.090	0.800	259	30.1	3.82	3.63	41.7	10.4	1.53
	53	15.6	12.250	7.870	1.090	0.620	216	24.1	3.72	3.28	38.5	9.80	1.57
ST12 x	50	14.7	12.000	7.245	0.870	0.745	215	26.3	3.83	3.84	23.8	6.58	1.27
	45	13.2	12.000	7.125	0.870	0.625	190	22.6	3.79	3.60	22.5	6.31	1.30
	40	11.7	12.000	7.000	0.870	0.500	162	18.7	3.72	3.29	21.1	6.04	1.34
ST10 x	48	14.1	10.150	7.200	0.920	0.800	143	20.3	3.18	3.13	25.1	6.97	1.33
	43	12.7	10.150	7.060	0.920	0.660	125	17.2	3.14	2.91	23.4	6.63	1.36
ST10 x	37.5	11.0	10.000	6.385	0.795	0.635	109	15.8	3.15	3.07	14.9	4.66	1.16
	33	9.70	10.000	6.255	0.795	0.505	93.1	12.9	3.10	2.81	13.8	4.43	1.19
ST9 x	35	10.3	9.000	6.251	0.691	0.711	84.7	14.0	2.87	2.94	12.1	3.86	1.08
	27.35	8.04	9.000	6.001	0.691	0.461	62.4	9.61	2.79	2.50	10.4	3.47	1.14
ST7.5 x	25	7.35	7.500	5.640	0.622	0.550	40.6	7.73	2.35	2.25	7.85	2.78	1.03
	21.45	6.31	7.500	5.501	0.622	0.411	33.0	6.00	2.29	2.01	7.19	2.61	1.07
ST6 x	25	7.35	6.000	5.477	0.659	0.687	25.2	6.05	1.85	1.84	7.85	2.87	1.03
	20.4	6.00	6.000	5.252	0.659	0.462	18.9	4.28	1.78	1.58	6.78	2.58	1.06
ST6 x	17.5	5.14	6.000	5.078	0.544	0.428	17.2	3.95	1.83	1.65	4.93	1.94	0.980
	15.9	4.68	6.000	5.000	0.544	0.350	14.9	3.31	1.78	1.51	4.68	1.87	1.00
ST5 x	17.5	5.15	5.000	4.944	0.491	0.594	12.5	3.63	1.56	1.56	4.18	1.69	0.901
	12.7	3.73	5.000	4.661	0.491	0.311	7.83	2.06	1.45	1.20	3.39	1.46	0.954
ST4 x	11.5	3.38	4.000	4.171	0.425	0.441	5.03	1.77	1.22	1.15	2.15	1.03	0.798
	9.2	2.70	4.000	4.001	0.425	0.271	3.51	1.15	1.14	0.941	1.86	0.932	0.831
ST3.5 x	7.65	2.25	3.500	3.662	0.392	0.252	2.19	0.816	0.987	0.817	1.32	0.720	0.766
ST3 x	8.625	2.53	3.000	3.565	0.359	0.465	2.13	1.02	0.917	0.914	1.15	0.648	0.675
	6.25	1.83	3.000	3.332	0.359	0.232	1.27	0.552	0.833	0.691	0.911	0.547	0.705

Properties shown in this table are for the full center split section.

Taken from Bethlehem Steel *Structural Shapes* handbook.

ANGLES
equal legs

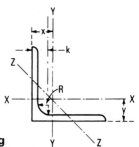

Theoretical Dimensions and Properties for **Designing**

Section Number and Size	Thickness	Weight per Foot	Area of Section	k	Axis X-X and Axis Y-Y				Axis Z-Z
					$I_{x,y}$	$S_{x,y}$	$r_{x,y}$	y or x	r_z
in.	in.	lb	in.²	in.	in.⁴	in.³	in.	in.	in.
L8 x 8 x	1⅛	56.9	16.7	1¾	98.0	17.5	2.42	2.41	1.56
R=⅝	1	51.0	15.0	1⅝	89.0	15.8	2.44	2.37	1.56
	⅞	45.0	13.2	1½	79.6	14.0	2.45	2.32	1.57
	¾	38.9	11.4	1⅜	69.7	12.2	2.47	2.28	1.58
	⅝	32.7	9.61	1¼	59.4	10.3	2.49	2.23	1.58
	9/16	29.6	8.68	1 3/16	54.1	9.34	2.50	2.21	1.59
	½	26.4	7.75	1⅛	48.6	8.36	2.50	2.19	1.59
L6 x 6 x	1	37.4	11.0	1½	35.5	8.57	1.80	1.86	1.17
R=½	⅞	33.1	9.73	1⅜	31.9	7.63	1.81	1.82	1.17
	¾	28.7	8.44	1¼	28.2	6.66	1.83	1.78	1.17
	⅝	24.2	7.11	1⅛	24.2	5.66	1.84	1.73	1.18
	9/16	21.9	6.43	1 1/16	22.1	5.14	1.85	1.71	1.18
	½	19.6	5.75	1	19.9	4.61	1.86	1.68	1.18
	7/16	17.2	5.06	15/16	17.7	4.08	1.87	1.66	1.19
	⅜	14.9	4.36	⅞	15.4	3.53	1.88	1.64	1.19
ʸL5 x 5 x	⅞	27.2	7.98	1⅜	17.8	5.17	1.49	1.57	0.973
R=½	¾	23.6	6.94	1¼	15.7	4.53	1.51	1.52	0.975
	ˣ ⅝	20.0	5.86	1⅛	13.6	3.86	1.52	1.48	0.978
	½	16.2	4.75	1	11.3	3.16	1.54	1.43	0.983
	7/16	14.3	4.18	15/16	10.0	2.79	1.55	1.41	0.986
	⅜	12.3	3.61	⅞	8.74	2.42	1.56	1.39	0.990
	5/16	10.3	3.03	13/16	7.42	2.04	1.57	1.37	0.994

ˣPrice—Subject to Inquiry.
ʸAvailability—Subject to Inquiry.

Taken from Bethlehem Steel *Structural Shapes* handbook.

ANGLES
unequal legs

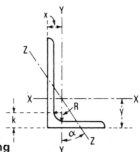

Theoretical Dimensions and Properties for **Designing**

Section Number and Size	Thickness	Weight per Foot	Area of Section	k	Axis X-X I_x	S_x	r_x	y	Axis Y-Y I_y	S_y	r_y	x	Axis Z-Z r_z	Tan α
in.	in.	lb	in.²	in.	in.⁴	in.³	in.	in.	in.⁴	in.³	in.	in.	in.	
L8 x 6 x	1	44.2	13.0	1½	80.8	15.1	2.49	2.65	38.8	8.92	1.73	1.65	1.28	0.543
R=½	⅞	39.1	11.5	1⅜	72.3	13.4	2.51	2.61	34.9	7.94	1.74	1.61	1.28	0.547
	¾	33.8	9.94	1¼	63.4	11.7	2.53	2.56	30.7	6.92	1.76	1.56	1.29	0.551
	9/16	25.7	7.56	1 1/16	49.3	8.95	2.55	2.50	24.0	5.34	1.78	1.50	1.30	0.556
	½	23.0	6.75	1	44.3	8.02	2.56	2.47	21.7	4.79	1.79	1.47	1.30	0.558
	7/16	20.2	5.93	15/16	39.2	7.07	2.57	2.45	19.3	4.23	1.80	1.45	1.31	0.640
L8 x 4 x		37.4	11.0	1½	69.6	14.1	2.52	3.05	11.6	3.94	1.03	1.05	0.846	0.247
R=½	⅞	33.1	9.73	1⅜	62.5	12.5	2.53	3.00	10.5	3.51	1.04	0.999	0.848	0.253
	¾	28.7	8.44	1¼	54.9	10.9	2.55	2.95	9.36	3.07	1.05	0.953	0.852	0.258
	⅝	24.2	7.11	1⅛	46.9	9.21	2.57	2.91	8.10	2.62	1.07	0.906	0.857	0.262
	9/16	21.9	6.43	1 1/16	42.8	8.35	2.58	2.88	7.43	2.38	1.07	0.882	0.861	0.265
	½	19.6	5.75	1	38.5	7.49	2.59	2.86	6.74	2.15	1.08	0.859	0.865	0.267
	7/16	17.2	5.06	15/16	34.1	6.60	2.60	2.83	6.02	1.90	1.09	0.835	0.869	0.269
L7 x 4 x	¾	26.2	7.69	1¼	37.8	8.42	2.22	2.51	9.05	3.03	1.09	1.01	0.860	0.324
R=½	⅝	22.1	6.48	1⅛	32.4	7.14	2.24	2.46	7.84	2.58	1.10	0.963	0.865	0.329
	½	17.9	5.25	1	26.7	5.81	2.25	2.42	6.53	2.12	1.11	0.917	0.872	0.335
	7/16	15.8	4.62	15/16	23.7	5.13	2.26	2.39	5.83	1.88	1.12	0.893	0.876	0.337
	⅜	13.6	3.98	⅞	20.6	4.44	2.27	2.37	5.10	1.63	1.13	0.870	0.880	0.340
L6 x 4 x	¾	23.6	6.94	1¼	24.5	6.25	1.88	2.08	8.68	2.97	1.12	1.08	0.860	0.428
R=½	⅝	20.0	5.86	1⅛	21.1	5.31	1.90	2.03	7.52	2.54	1.13	1.03	0.864	0.435
	9/16	18.1	5.31	1 1/16	19.3	4.83	1.90	2.01	6.91	2.31	1.14	1.01	0.866	0.438
	½	16.2	4.75	1	17.4	4.33	1.91	1.99	6.27	2.08	1.15	0.987	0.870	0.440
	7/16	14.3	4.18	15/16	15.5	3.83	1.92	1.96	5.60	1.85	1.16	0.964	0.873	0.443
	⅜	12.3	3.61	⅞	13.5	3.32	1.93	1.94	4.90	1.60	1.17	0.941	0.877	0.446
	5/16	10.3	3.03	13/16	11.4	2.79	1.94	1.92	4.18	1.35	1.17	0.918	0.882	0.448
ʸ**L6 x 3½ x**	⅜	11.7	3.42	⅞	12.9	3.24	1.94	2.04	3.34	1.23	0.988	0.787	0.767	0.350
R=½	5/16	9.8	2.87	13/16	10.9	2.73	1.95	2.01	2.85	1.04	0.996	0.763	0.772	0.352

ʸAvailability—Subject to Inquiry.

Taken from Bethlehem Steel *Structural Shapes* handbook.

TABLE A STEEL BEAMS DATA

American Standard Channel

H-Pile

Angle: Equal Legs

Angle: Unequal Legs

American Standard

American Standard Tee

Wide Flange

Wide Flange Tee

TABLE B — Steel Beam Clips Data

Size and length	6 x 4 x 3/8 in 0 ft 2 in long
Weight inc. web rivets	5 lb
Connection details	(2 3/4", 2 3/4", g, g; 2 1/4", 2 1/2", A; 1 3/4")
Nominal Beam/Column Size	3 in, 4 in Beam / No min. Col. Size
Size and length	6 x 4 x 3/8 in 0 ft 2 1/2 in long
Weight inc. web rivets	6 lb
Connection details	(2 3/4", 2 3/4", g, g; 2 1/4", 2 1/2", A; 1 3/4")
Nominal Beam/Column Size	5 in, 6 in Beam / No min. Col. Size

TABLE B — STEEL BEAM CLIPS DATA

Size and length	6 x 4 x 3/8 in 0 ft 5 1/2 in long
Weight inc. web rivets	13 lb
Connection details	(2 3/4", 2 3/4", g, g; 2 1/4", 2 1/2", A, 3", 1 3/4")
Nominal Beam/Column Size	8in, 10in, 12in BEAM 10in column 30lb and under 8in column 30lb. and under 12in column 50lb. and under
Size and length	6 x 6 x 3/8 in 0 ft 5 1/2 in long
Weight inc. web rivets	16 lb
Connection details	(2 1/4", 2 3/4", 2 3/4", 2 1/4", g, g; 2 1/2", 2 1/4", A, 3", 2")
Nominal Beam/Column Size	8in, 10in, BEAM 10in column over 30lb 8in column over 30lb

TABLE B — Steel Beam Clips Data

Size and length	4 x 3 1/2 x 3/8 in 0 ft 11 1/2 in long
Weight inc. web rivets	19 lb
Connection details	*(diagram: beam web connection with 2 3/4" g dimensions; column connection with 2 1/4", 3", 3", 3", 1 3/4" dimensions, dimension A)*
Nominal Beam/Column Size	14in, 16in, 18in BEAM 16in column 50lb and under 14in column 42lb. and under
Size and length	6 x 6 x 3/8 in 0 ft 8 1/2 in long
Weight inc. web rivets	24 lb
Connection details	*(diagram: beam web connection with 2 1/4", 2 3/4", 2 3/4", 2 1/4" g dimensions; column connection with 2 1/2", 2 1/4", 3", 3", 2" dimensions, dimension A)*
Nominal Beam/Column Size	12in BEAM 12in column over 50lb 14in column over 42lb

TABLE B

STEEL BEAM CLIPS DATA

Size and length	6 x 6 x 3/8 in 0 ft 11 1/2 in long
Weight inc. web rivets	33 lb
Connection details	
Nominal Beam/Column Size	16in BEAM 16in column over 50lb
Size and length	6 x 6 x 3/8 in 1 ft 2 1/2 in long
Weight inc. web rivets	41 lb
Connection details	
Nominal Beam/Column Size	18in, 20in BEAM 18in column over 58lb

TABLE C

Species and Commerical Grade	Douglas Fir Structural Data Compression Parallel to Grain	Modulus of Elasticity
	F_c (psi)	E (psi)
Dimension Lumber 2 to 4 in. thick, 2in. and wider (Moisture not exceeding 19%)		
Select Structure	1700	1,900,000
No. 1 and Better	1500	1,800,000
No. 1	1300	1,600,000
No. 2	1300	1,600,000
No. 3	750	1,400,000
Stud	825	1,400,000
Construction	1600	1,500,000
Standard	1350	1,400,000
Utility	875	1,300,000

Source: National Design Specification for Wood Construction, 1991 edition (Ref. 5) Values listed are for normal duration loading with wood that is surfaced dry or green and used at 19% maximum moisture content. Excerpt from Parker and Ambrose, *Simplified Engineering for Architects and Builders*, 8th ed., 1993. Reprinted by permission of John Wiley & Sons, Inc.

TABLE D

CONCRETE REINFORCING BAR DATA

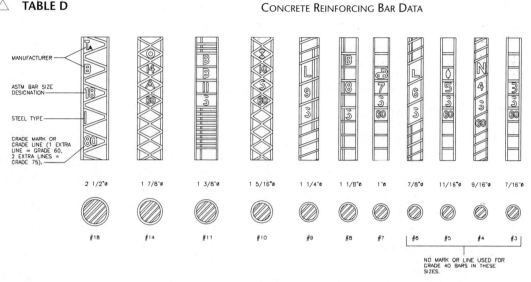

REINFORCING STEEL GRADES AND STRENGTHS

ASTM SPECIFICATION	MINIMUM YIELD STRENGTH (PSI)	MINIMUM TENSILE STRENGTH (PSI)	STEEL TYPE
BILLET: ASTM A-615			
GRADE 40	40,000	70,000	
GRADE 60	60,000	90,000	S
GRADE 75	75,000	100,000	
RAIL: ASTM A-616			
GRADE 50	50,000	80,000	I
GRADE 60	60,000	90,000	
AXIL: ASTM A-617			
GRADE 40	40,000	70,000	A
GRADE 60	60,000	90,000	
LOW-ALLOY: ASTM A-706			
GRADE 60	60,000	80,000	W

Adapted from *Architectural Graphic Standards*, 8th ed., by Ramsey/Sleeper. John Wiley & Sons, Inc. publishers. Copyright 1994. Reprinted by permission of John Wiley & Sons, Inc.

△ △ **TABLE E1**

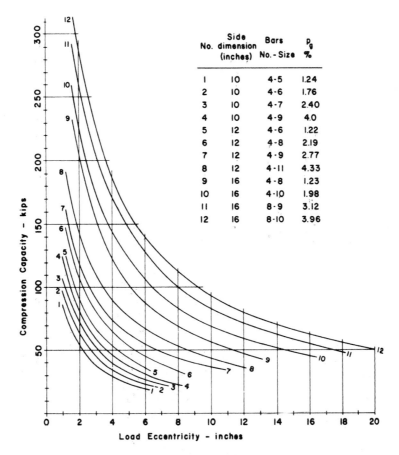

Adapted from *Simplified Engineering for Architects and Builders,* 8th ed., Ramsey/Sleeper, 1994. John Wiley & Sons, Inc. publishers. Copyright 1994. Reprinted by permission of John Wiley & Sons, Inc.

TABLE E2

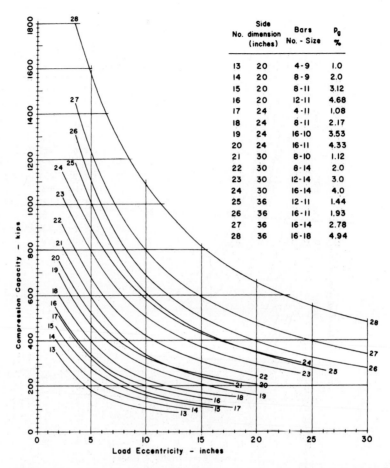

Adapted from *Simplified Engineering for Architects and Builders,* 8th ed., Ramsey/Sleeper, 1994. John Wiley & Sons, Inc. publishers. Copyright 1994. Reprinted by permission of John Wiley & Sons, Inc.

△ △ **TABLE E3**

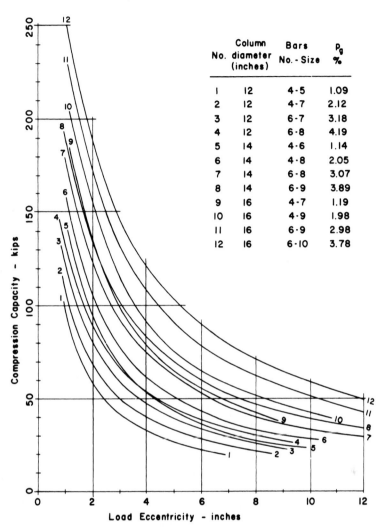

No.	Column diameter (inches)	Bars No.-Size	p_g %
1	12	4-5	1.09
2	12	4-7	2.12
3	12	6-7	3.18
4	12	6-8	4.19
5	14	4-6	1.14
6	14	4-8	2.05
7	14	6-8	3.07
8	14	6-9	3.89
9	16	4-7	1.19
10	16	4-9	1.98
11	16	6-9	2.98
12	16	6-10	3.78

Adapted from *Simplified Engineering for Architects and Builders,* 8th ed., Ramsey/Sleeper, 1994. John Wiley & Sons, Inc. publishers. Copyright 1994. Reprinted by permission of John Wiley & Sons, Inc.

TABLE E4

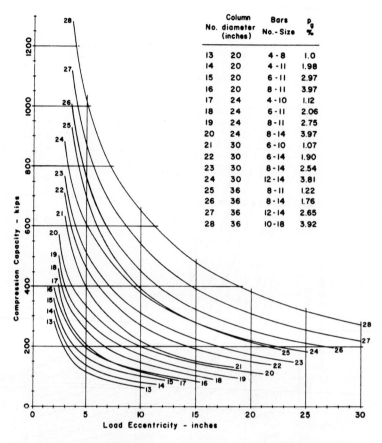

Adapted from *Simplified Engineering for Architects and Builders,* 8th ed., Ramsey/Sleeper, 1994. John Wiley & Sons, Inc. publishers. Copyright 1994. Reprinted by permission of John Wiley & Sons, Inc.

TABLE F1 — SQUARE COLUMN FOOTINGS

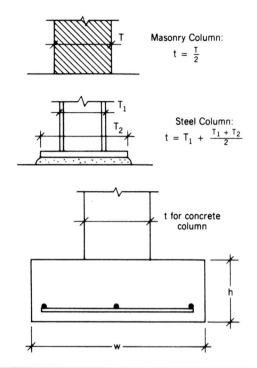

Masonry Column:
$t = \frac{T}{2}$

Steel Column:
$t = T_1 + \frac{T_1 + T_2}{2}$

t for concrete column

Maximum Soil Pressure (lb/ft^2)	Minimum Column Width t (in.)	$f'_c = 2000$ psi					$f'_c = 3000$ psi			
		Allowable Load^a on Footing (k)	Footing Dimensions h (in.)	Footing Dimensions w (ft)	Reinforcing Each Way		Allowable Load^a on Footings (k)	Footing Dimensions h (in.)	Footing Dimensions w (ft)	Reinforcing Each Way
1000	8	7.9	10	3.0	2 No. 3		7.9	10	3.0	2 No. 3
	8	10.7	10	3.5	3 No. 3		10.7	10	3.5	3 No. 3
	8	14.0	10	4.0	3 No. 4		14.0	10	4.0	3 No. 4
	8	17.7	10	4.5	4 No. 4		17.7	10	4.5	4 No. 4
	8	22	10	5.0	4 No. 5		22	10	5.0	4 No. 5
	8	31	10	6.0	5 No. 6		31	10	6.0	5 No. 6
	8	42	12	7.0	6 No. 6		42	11	7.0	7 No. 6
1500	8	12.4	10	3.0	3 No. 3		12.4	10	3.0	3 No. 3
	8	16.8	10	3.5	3 No. 4		16.8	10	3.5	3 No. 4
	8	22	10	4.0	4 No. 4		22	10	4.0	4 No. 4

Adapted from *Simplified Engineering for Architects and Builders,* 8th ed., Ramsey/Sleeper, 1994. John Wiley & Sons, Inc. publishers. Copyright 1994. Reprinted by permission of John Wiley & Sons, Inc.

TABLE F1 (Continued)

Maximum Soil Pressure (lb/ft²)	Minimum Column Width t (in.)	f'_c = 2000 psi				f'_c = 3000 psi			
		Allowable Load[a] on Footing (k)	Footing Dimensions h (in.)	Footing Dimensions w (ft)	Reinforcing Each Way	Allowable Load[a] on Footing (k)	Footing Dimensions h (in.)	Footing Dimensions w (ft)	Reinforcing Each Way
	8	28	10	4.5	4 No. 5	28	10	4.5	4 No. 5
	8	34	11	5.0	5 No. 5	34	10	5.0	6 No. 5
	8	48	12	6.0	6 No. 6	49	11	6.0	6 No. 6
	8	65	14	7.0	7 No. 6	65	13	7.0	6 No. 7
	8	83	16	8.0	7 No. 7	84	15	8.0	7 No. 7
	8	103	18	9.0	8 No. 7	105	16	9.0	10 No. 7
2000	8	17	10	3.0	4 No. 3	17	10	3.0	4 No. 3
	8	23	10	3.5	4 No. 4	23	10	3.5	4 No. 4
	8	30	10	4.0	6 No. 4	30	10	4.0	6 No. 4
	8	37	11	4.5	5 No. 5	38	10	4.5	6 No. 5
	8	46	12	5.0	6 No. 5	46	11	5.0	5 No. 6
	8	65	14	6.0	6 No. 6	66	13	6.0	7 No. 6
	8	88	16	7.0	8 No. 6	89	15	7.0	7 No. 7
	8	113	18	8.0	8 No. 7	114	17	8.0	9 No. 7
	8	142	20	9.0	8 No. 8	143	19	9.0	8 No. 8
	10	174	21	10.0	9 No. 8	175	20	10.0	10 No. 8
3000	8	26	10	3.0	3 No. 4	26	10	3.0	3 No. 4
	8	35	10	3.5	4 No. 5	35	10	3.5	4 No. 5
	8	45	12	4.0	4 No. 5	46	11	4.0	5 No. 5
	8	57	13	4.5	6 No. 5	57	12	4.5	6 No. 5
	8	70	14	5.0	5 No. 6	71	13	5.0	6 No. 6
	8	100	17	6.0	7 No. 6	101	15	6.0	8 No. 6
	10	135	19	7.0	7 No. 7	136	18	7.0	8 No. 7
	10	175	21	8.0	10 No. 7	177	19	8.0	8 No. 8
	12	219	23	9.0	9 No. 8	221	21	9.0	10 No. 8
	12	269	25	10.0	11 No. 8	271	23	10.0	10 No. 9
	12	320	28	11.0	11 No. 9	323	26	11.0	12 No. 9
	14	378	30	12.0	12 No. 9	381	28	12.0	11 No. 10
4000	8	35	10	3.0	4 No. 4	35	10	3.0	4 No. 4
	8	47	12	3.5	4 No. 5	47	11	3.5	4 No. 5
	8	61	13	4.0	5 No. 5	61	12	4.0	6 No. 5
	8	77	15	4.5	5 No. 6	77	13	4.5	6 No. 6
	8	95	16	5.0	6 No. 6	95	15	5.0	6 No. 6
	8	135	19	6.0	8 No. 6	136	18	6.0	7 No. 7
	10	182	22	7.0	8 No. 7	184	20	7.0	9 No. 7
	10	237	24	8.0	9 No. 8	238	22	8.0	9 No. 8
	12	297	26	9.0	10 No. 8	299	24	9.0	9 No. 9
	12	364	29	10.0	13 No. 8	366	27	10.0	11 No. 9
	14	435	32	11.0	12 No. 9	440	29	11.0	11 No. 10
	14	515	34	12.0	14 No. 9	520	31	12.0	13 No. 10
	16	600	36	13.0	17 No. 9	606	33	13.0	15 No. 10
	16	688	39	14.0	15 No. 10	696	36	14.0	14 No. 11
	18	784	41	15.0	17 No. 10	793	38	15.0	16 No. 11

[a] *Note:* Allowable loads do not include the weight of the footing, which has been deducted from the total bearing capacity. Criteria: f_s = 20 ksi, $v_c = 1.1\sqrt{f'_c}$ for beam shear, $v_c = 2\sqrt{f'_c}$ for peripheral shear.

Adapted from *Simplified Engineering for Architects and Builders,* 8th ed., Ramsey/Sleeper, 1994. John Wiley & Sons, Inc. publishers. Copyright 1994. Reprinted by permission of John Wiley & Sons, Inc.

TABLE F2 — WALL FOOTINGS

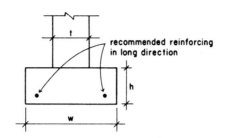

Maximum Soil Pressure (lb/ft²)	Minimum Wall Thickness		Allowable Load[a] on Footing (lb/ft)	Footing Dimensions		Reinforcing	
	Concrete t (in.)	Masonry t (in.)		h (in.)	w (in.)	Long Direction	Short Direction
1000	4	8	2,625	10	36	3 No. 4	No. 3 at 16
	4	8	3,062	10	42	2 No. 5	No. 3 at 12
	6	12	3,500	10	48	4 No. 4	No. 4 at 16
	6	12	3,938	10	54	3 No. 5	No. 4 at 13
	6	12	4,375	10	60	3 No. 5	No. 4 at 10
	6	12	5,250	10	72	4 No. 5	No. 5 at 11
1500	4	8	4,125	10	36	3 No. 4	No. 3 at 10
	4	8	4,812	10	42	2 No. 5	No. 4 at 13
	6	12	5,500	10	48	4 No. 4	No. 4 at 11
	6	12	6,131	11	54	3 No. 5	No. 5 at 15
	6	12	6,812	11	60	5 No. 4	No. 5 at 12
	8	16	8,100	12	72	5 No. 5	No. 5 at 10
2000	4	8	5,625	10	36	3 No. 4	No. 4 at 14
	6	12	6,562	10	42	2 No. 5	No. 4 at 11
	6	12	7,500	10	48	4 No. 4	No. 5 at 12
	6	12	8,381	11	54	3 No. 5	No. 5 at 11
	6	12	9,250	12	60	4 No. 5	No. 5 at 10
	8	16	10,875	15	72	6 No. 5	No. 5 at 9
3000	6	12	8,625	10	36	3 No. 4	No. 4 at 10
	6	12	10,019	11	42	4 No. 4	No. 5 at 13
	6	12	11,400	12	48	3 No. 5	No. 5 at 10
	6	12	12,712	14	54	6 No. 4	No. 5 at 10
	8	16	14,062	15	60	5 No. 5	No. 5 at 9
	8	16	16,725	17	72	6 No. 5	No. 6 at 10

[a] *Note:* Allowable loads do not include the weight of the footing, which has been deducted from the total bearing capacity. Criteria: $f'_c = 2000$ psi, $f_s = 20$ ksi, $v_c = 1.1\sqrt{f'_c}$.

Adapted from *Simplified Engineering for Architects and Builders,* 8th ed., Ramsey/Sleeper, 1994. John Wiley & Sons, Inc. publishers. Copyright 1994. Reprinted by permission of John Wiley & Sons, Inc.

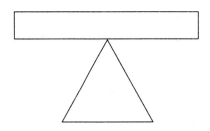

Glossary

A

Acting forces—Forces acting upon a system that tend to move the system out of static equilibrium.

Actual—The true dimensions of a member after it has been finished to specifications.

Acute—Angles that have a value between 0 and 90°.

Additive—Colinear forces acting in the same direction.

ANSI—American National Standards Institute.

ASTM—American Society for Testing and Materials.

Axial compression—Also known as *axial load*. Any load that acts parallel to the long axis of a column.

B

Beam—Horizontal structural support member.

Beam clips—Segments of steel angles used to fasten steel structural elements together.

Beam marks—Used to tell the manufacturer how to mark the beam before it is sent to the construction site.

Beam reaction—A stabilizing force that is a direct response by the beam itself to forces applied to it.

Bending moment—The amount of work required to bend a column or the amount of work being done on a column that tends to cause bending.

Block—Similar to a cope with an additional portion of the web removed.

Block diagram—Each linear foot of an object is represented by a single block.

Board-foot—A unit of lumber volume. The computed value of one-twelfth of the length of lumber in feet multiplied by the breadth times depth in inches.

Breaking strength—The ability of an object to resist being broken across an axis.

Buckling—A type of defomation caused by "weak points" located along the length of a slender object, such as a post or column.

C

Centroid—The center of mass of an object.

Clip dimensions—Contain the clip nomenclature, including information on how many sides of the beam the clips are located.

Clip list—A list that contains all of the information about each clip used in a structure.

Coincident—Two or more forces that all act upon the same point.

Colinear—Two or more forces that act along the same line of action.

Column line dimensions—Shows the center-to-center distance between columns.

Column lines—Lines used to indicate the center lines of the columns on the plan.

Columns—Vertical support members of a structure.

Component form—A vector that is expressed using components.

Components—A single vector represented by the two legs of a right triangle.

Compression—A force that tends to cause the shortening of an object along an axis.

Compressive strength—The ability of an object to resist being shortened along an axis.

Concrete formation—Indicates whether the concrete is field-formed, pre-cast, or formed under specific conditions. Also may indicate limitations, such as pouring temperature and setting time.

Concrete strength—The compressive or yield strength of concrete measured in pounds per square inch.

Concrete type—Indicates the material composition of concrete used to form a structure.

Conductivity—The ability of a substance to transfer heat and/or electricity. One of the properties of metals.

Cope—A beam cut in which the flange of the beam is removed down to the web at the point where the fillet meets the flat portion of the web.

Coplanar—Two or more forces that act within the same plane.

Cut—Removes only a portion of the beam flange and does not affect the web at all.

Cut and chip—Almost identical to a cut except that the inset is equal to the a value of the beam and is not dependent on any of the column dimensions.

D

Decking—The surface of the flooring that rests on the joists.

Deflection—The vertical deviation of a beam over a horizontal span.

Deformation—The change that occurs to the shape and size of an object due to the forces acting upon it.

Direction—The angle commonly measured from the horizontal or from a chosen line at which a force is applied. The angle is measured counterclockwise from the positive x axis to the line of action.

Double shear—Occurs when shear acts at two points along a rivet. This value is twice the single shear value.

Ductility—The ability of a substance to be drawn out, extruded, or folded into a new shape. One of the properties of metals.

E

Eccentricity—The imaginary distance from a column's center axis at which an axial load (P) is considered to be placed in order to generate a bending moment.

Effective falling distance—The approximate distance that an object would need to fall to reach a given velocity.

Elastic limit—The load that is required to reach the point of permanent deformation of a material.

End joist—Joist that runs outside of the floor frame parallel to the floor joists.

External beam forces—All forces applied to a beam system from outside the beam.

External forces—Forces acting upon a system that originate outside the system itself.

F

Failure—The point at which deformation permanently compromises the structural integrity of an object.

Fatigue—The weakening of an object due to deformation or repeated loading and unloading.

Flexure—The bending of a beam that causes compression of extreme fibers on the side that a force is applied and tension of extreme fibers on the side opposite to the force.

Flooring—The materials used to construct a floor. Consists mainly of joists and decking.

Footing—The lowest member of a foundation system. Used to support and distribute loads from a structure across the supporting soil.

Foot-pound (ft-lb)—The amount of work required to raise a one-pound object to a height of one foot above its original position.

Force—A motivator that acts along a specific line in a specific direction, which tends to bring about a change in the state of a system.

Force at impact—The maximum reading on a gauge when an object strikes a surface. An instantaneous force.

Force triangle—A diagram that can also be constructed to show the relationship between components and a given vector.

Fulcrum—The fixed point on a moment arm. Also known as the *point of interest*.

G

Gage distance—The distance from the back of the clip to the centerline of the clip holes.

Gage hole dimensions—The number of rivet or bolt holes, their spacing, and total O.C. length along the mount side of the clip.

Gage line—The centerline that runs in any direction through a hole or row of holes parallel to any one of the planes of the clip.

Glue-laminated lumber—Wood laminated with an epoxy. Stronger and smoother than hand-finished lumber.

H

Header joist—Joist that runs outside of the floor frame perpendicular to the floor joists.

Heat-tempered steel—Steel that is formed by quickly cooling a heated piece of standard steel. The resulting metal is more rigid, but tends to be more brittle than standard steel.

High-carbon steel (carbide)—Steel made with a higher concentration of carbon. This form of steel takes on a blue-black or black color. It is less dense than standard steel with a much higher compressive strength.

Hooke's law—The law that states that an object will experience a uniform rate of deformation under uniformly increasing stresses.

Horizontal—Parallel to the x axis, $(+)$ = right, $(-)$ = left. This is level with the horizon.

I

Independent footing—A footing designed to support a single post or column and its accompanying load.

Internal beam forces—All forces on a beam system due solely to the weight of the beam.

Internal forces—Forces that act within a system due to the weight of the members that make up that system.

J

Joist—Horizontal structural support member spaced at regular intervals under flooring or decking.

K

Kinetic energy—A type of imparted (released) energy that is measured as a function of the velocity of an object at impact with a surface.

L

Law of conservation of energy—States that energy can neither be created nor destroyed, but can be transformed.

Line of force—The line that represents the direction of any single force.

LOA (line of action)—The line along which an external force is applied to a system.

LOI (line of interest)—The line upon which an external force is applied to a system.

M

Magnitude—The numerical value of the intensity of a force; usually measured in pounds, kilopounds (kips), tons (2000 pounds), kilograms, or similar units.

Malleability—The ability of a substance to be struck and reshaped by a hammer.

Maximum material condition—The dimension(s) that will result in the largest area of a cross section of a material.

Milling—The specification by which the dimensions of a wood member are reduced. Used to determine the actual size of a piece of lumber.

Minimum material condition—Also known as *least material condition* or *LMC*. The dimension(s) that will result in the smallest area of a cross section of material.

Miscellaneous dimensions—Used to show detailed dimensioning of any features not called out by any of the other three dimensioning methods.

Modulus of elasticity—A property that reveals the stiffness of a material.

Moment—The effect of a force applied to an object at a distance from a fixed point, which causes rotation about that fixed point.

Moment of inertia—A calculated value based on the shape of the cross section of an object that reveals the strength of that shape.

Multiple force/constant distance—One of the work scenarios. In this scenario two or more forces act along the same distance.

Multiple force/multiple distance—One of the work scenarios. The sum of several forces acting along varying distances.

N

Negating—Colinear forces acting in opposite directions.

Newton's laws of motion

First—Inertia. An object at rest tends to stay at rest and an object in motion tends to remain in constant motion in a straight line unless acted upon by an outside force.

Second—Momentum. External forces acting upon an object cause that object to accelerate uniformly in direct proportion to the force being applied.

Third—For every action there exists an equal and opposite reaction.

Nominal—Sizes that are used for element nomenclature but are not necessarily the true dimensions of the member.

Normal Force—The component of a force that is perpendicular to a given surface.

O

Overall dimensions—Show the distance between the outermost features of the plan.

Oxidation—The combination of oxygen with a substance. Causes rust or slag in metals.

Oxidize—To experience oxidation. Steel is more resistant to oxidation than iron.

P

Permanent set—When an object retains a deformed shape after an external stress has been removed.

Pitch—The distance between gage lines measured along the length of the clip.

Pictorial diagram—A sketch that represents a word problem in picture form.

POA (point of action)—The point along a line of interest at which an external force is applied.

Point–force diagram—A diagram on which all of the lines, angles, forces, and references are labeled.

POI (point of interest)—The fixed point of a moment arm about which rotation occurs. Also known as the *fulcrum*.

Post—Also called a *column*. Vertical structural support member.

Potential energy—A type of stored energy. Measured as a function of an object's position relative to the ground.

Primary structural member—A member that is directly attached to the foundation by a column or columns.

R

Reacting forces—The stabilizing forces that respond to acting forces to keep a system in static equilibrium.

Rebar—Steel reinforcing bar. Used to enhance the structural integrity of concrete.

Resultant—The vector representation that shows the magnitude and direction of the vector.

Round lumber—Lumber that has been cut into a round section, usually using a lathe.

S

Sawn lumber—Lumber that has been cut and is unfinished.

Secondary structural member—A member that does not directly transfer the load it carries to the foundation of the structure.

Shear—A force that acts at a sharp angle to an object and tends to cause that object to break across an axis.

Single force/constant distance—One of the work scenarios. In this scenario a single object moves through a single distance.

Single force/multiple distance—One of the work scenarios. In this scenario a single object moves through more than one distance.

Single shear—Occurs when two surfaces that have forces acting in opposite directions are riveted together.

Solid sawn—A member that has been cut and shaped from a single piece of wood.

Spreader—A steel beam that spans two other horizontal members or beams. Used to laterally separate and reinforce beams.

Spreader/joist spacing—This dimension is called out as a note over a double-sided arrow and indicates the standard spacing on-center between spreaders or joists.

Stabilizing—A force that has the same magnitude as the sum of the acting forces on a system, but acts in the opposite direction.

Static equilibrium—Occurs when the sum of the forces acting on a body produces no motion in the body, and the body remains at rest.

Strength of materials—Also called *mechanics of materials*. The study of the properties of the materials of bodies that enable them to resist the actions, stresses, and deformations caused by external forces.

Structural analysis—The study of the materials, type(s), and purpose(s) of a structure along with an investigation of the internal and external forces applied to the system structure.

Structural integrity—The ability of an object or system to resist forces internally and externally applied to that system.

Subflooring—The part of the decking that rests directly on the joists.

T

Tensile strength—The ability of an object to resist being lengthened along an axis.

Tension—A force that tends to cause an object to lengthen along an axis.

Thermal shock—The effect that a sudden change in temperature has on a material.

Torsion—Also called *torque*. A force that causes an object to twist or rotate about an axis.

Torsional strength—The ability of an object to resist being twisted about an axis.

Total load—The sum of all live and dead loads.

V

Vector—A line drawn to represent a force.

Velocity—The rate of change of displacement of an object with respect to time. Measured in miles per hour (mph), feet per second (fps), etc.

Vertical—Parallel to the y axis, $(+) =$ up, $(-) =$ down. This is perpendicular to the horizon.

Visually graded lumber—Hand-finished lumber without the use of fine machine measurements.

W

Wall footing—A footing that supports the entire length of a load-bearing wall.

Work—A force acting through a distance. Measured in foot-pounds (ft-lb).

Y

Yield point—The physical point at which the elastic limit of an object is reached; that is, the point beyond which an object will no longer deform uniformly.

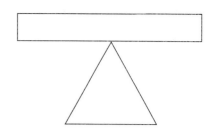

Bibliography

1. Ramsey/Sleeper. *Architectural Graphic Standards*, 8th ed. New York: Wiley, 1994.
2. BOCA. *The BOCA National Building Code 1993*. BOCA International, 1993.
3. Halliday/Resnick. *Fundamentals of Physics* New York: Wiley, 1981.
4. Boyce/Margolis/Slade. *Mathematics for Technical and Vocational Students*. Englewood Cliffs, NJ: Prentice Hall, 1989.
5. Parker/Ambrose. *Simplified Engineering for Architects and Builders*, 8th edition. New York: Wiley, 1993.
6. Green, Robert E. *Machinery's Handbook*, 24th ed. New York: Industrial Press, 1992.
7. Bethlehem Steel. *Structural Shapes*. Bethlehem, PA: Bethlehem Steel Corp., 1989.
8. Dull, Charles E. *Modern Physics*. Henry Holt, 1934.
9. Putnam, Robert. Builder's Comprehensive Dictionary. Reston, VA: Craftsman Book, 1989.
10. *Encyclopedia Britannica*, 15th ed. Chicago, Il: Encyclopedia Britannica, 1985, Vol. 21, pp. 418–419.

ADDITIONAL READING

The Principles of Building Construction, by Madan Mehta. Prentice Hall, 1997.

Applied Mechanics for Engineering Technology, 5th ed., by Keith M. Waller. Prentice Hall, 1997.

Building Construction—Principles, Practices, and Materials, by Glenn M. Hardie. Prentice Hall, 1995.

Applied Strengths of Materials, 3rd ed., by Robert L. Mott. Prentice Hall, 1996.

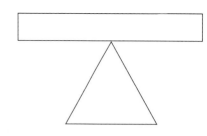

Index*

Acting forces (*see* Forces, acting)
Additive (forces) (*see* Forces, colinear)
Analysis (structural) (*see* Structures, analysis of)
ASTM, **126**

Beams:
    flexure in, **124**
    forces (*see* Forces, beam)
    reactions, **94**, 97
    sizing, 94
    steel (*see* Steel, beams)
    wood (*see* Wood, beams and joists)
Bending moments (*see* Moments, bending)
Block (*see* Steel, beams, custom cuts)
Block diagrams, **100**
Bolts (steel) (*see* Steel, beams, fasteners)
Breaking strength (*see* Reactions, as material strengths, breaking)
Buckling (*see* Columns, buckling in)

Centroids, **67**, 69, 127
    location of in beams, 68
    of circles, 180, 182
    of triangles, 180, 181
    of rectangles, 127, 181
Clips (steel) (*see* Steel, beams, fasteners, *see* Steel, clips)

Columns (*see also* Concrete, reinforced columns)
    (*see also* Wood, columns and posts) 99, 130, 132, 135, 185, 198
    buckling in, 212
    compressions on, **99**, 101, **198**, 212–13
Compression (*see* Forces, types, compression)
Compressive strength (*see* Reactions, as material strengths, compressive)
Concrete:
    detailing of, 223–25
        criteria, 220, 223–25
    footings, **219**
        independent, **219**
        wall, **220**
    rebar, **209**
        nomenclature, 210–11
        placement of, 224, 225
    reinforced columns, 215–17
Cope (*see* Steel, beams, custom cuts)
Cut (*see* Steel, beams, custom cuts)
Cut and chip (*see* Steel, beams, custom cuts)

Deflection, 113, **123**, 124, 126, 139
    basic formula, 123
    standard formula, 139
Deformation, 120, **123**

---

*This index contains a comprehensive listing of all technical terms used in this text. References are given for each new appearance of a word or term as well as the definition of that term. The page number(s) on which the definition(s) of a term may be found are indicated in this index using bold numbers. Reference will only be made to the glossary (pages prefixed with a G) if the definition of a term is not found within the main body of the text.

315

Eccentricity, 212, **214**, 215, 217
   units of, 214
Effective falling distance, 49, **G307**
Elastic limit, **124**
Energy, 83
   due to velocity, 50, 51, 52
   gain/loss, 46, 48
   kinetic, **49**, 50
   positional, 46–49
   potential, **46**, 47–49
   units of, 39, 47–48

Failure, 120, **124**
Fatigue, 120, **124**
Flexure (*see* Beams, flexure in)
Footings (concrete), 219 (*see also* Concrete, footings)
Force, **5**
   line of, **9**
   triangle, **18**, 19
Forces:
   acting, 6, 9, 10, 68, **G305**
   beam, 93
      external, **93**
      internal, **93**, 97
   coincident, **9**
   colinear, **8**
      additive, **9**
      negating, **9**
   components of, **18**, 77, 78
   coplanar, **9**
   direction of, **5**, 62–63, 96
   external, **9**, **65**, 68, 93, 103
   internal, **9**, **93**, 101
   magnitude of, **5**, 77
   normal, **76**, 77–81
   sense, 10
   reacting (stabilizing), 7, **10**, **25**, 72, 73, 80, 94, **G311**
   types:
      compression, **6**
      shear, **6**
      tension, **6**
      torsion (torque), **6**, 65
Fulcrum, **61**, 62, 63, 64, 70 (*see also* Moments, point of, interest)

Gravity, constant of acceleration: **50**

Hooke's law, **124**

I-beams (*see* Steel, beams)
Inertia (*see* Newton's laws of motion)
Integrity (*see* Structures, integrity of)

Joist (wood) (*see* Wood, joists)

Kinetic energy (*see* Energy, kinetic)

Laws of
   conservation of energy, **49**
   motion (*see* Newton's laws of motion)
Least material condition, **185**
Levers:
   first order, **64**, 70, 73
   second order, **70**, 74
Loads:
   axial on columns, **213**
   dead, 124
   distribution of, 127–32
      concentric, 212
      nonconcentric, 217
      nonuniform, **97**, 100
      radial, 212
      uniform, 97, 101, 127
   live, 124
   total, **198**
Lumber (*see also* Wood):
   board-feet, **187**, 202
   glue-laminated, 198
   grade, 188
   square, 181, 199, 201
   round, 182, 191–92, **199**
   solid sawn, **198**
   visually graded, **198**

Maximum material condition, **G309**
Metal, properties of, 121
   conductivity, **121**
   ductility, **121**
   malleability, **121**
Minimum material condition (*see* Least material condition)
Modulus of elasticity, **125**
Moments, **61**, 95
   arm (*see* line of, interest)
   bending, 123, **213**, 214, 215
   convention of, 62, 63
   line of:
      action, 62
      interest, 61
   point of:
      action, **61**
      interest, **61**
   sum of, 66, 69, 71
   units of, 61, 65
Moment of inertia, **179** (*see also* Steel, beams, sizing)

Newton's laws of motion, 94 (third), **G310**

Normal (forces) (*see* Forces, normal)

Oxidation, **119**, 125

Pictorial diagrams, **20**
Permanent Set, **124**
Point–force diagrams (*see* Vectors, point–force diagrams) (*see* Block diagrams)
Post (*see* Wood, columns and posts)
Post and beam, 205
Potential energy (*see* Energy, potential)

Reactions:
    as material strengths (*see also* Strength of materials):
        breaking, **7**, 51, 52, 53, 55
        compressive, **7**
        tensile, **7**
        torsional, **7**
    due to actions (*see* Newton's laws of motion)
    on beams (*see* Beams, reactions)
Rebar (*see* Concrete, rebar)
Resultant (*see* Vectors, resultant form)
Rivets (steel) (*see* Steel, beams, fasteners)

Safety factor, 159–60
Shear (*see also* Forces, types, shear):
    double, **158**
    in bolts, 157
    in rivets, 157
    single, **157**
Static equilibrium, **1**, 5, 24–25, **66**, 96
    conditions of, **1**, 25, 53, 82, 107
Steel, **119**
    architectural, 122–25
    beams:
        allowable deflection, 124
        custom cuts, 164–68, 169, 170
        cutting lengths, 135–37, **138**
        cutting lists, 169
        detail, **169**, 170
        dimensions, 134–35
        fasteners, 157–58
        joists, **153**, 154–55
        load-capacity calculations, 141–42
        moment of inertia in, 139, 140
        nomenclature, 132, 133
        orientations of, 136, 137, 138
        parts of, 134–35
        shapes of, 132, **133**, 134
        selection, 140–41, 156
        sizing, 139–41, 153–55

        clips (beam), 155, 161
            detailing of, 161–63
            dimensions, **161**
            gage, **161**
            list, **162**, 163
            pitch, **161**
            selection, 155–56
    framing plan, 127, 152–53
        beam marks, **153**
        column lines, **153**
    grades, 211
    manufacturing methods, 121, 122, 125
    modulus of elasticity, **125**, 211
    stainless, 119
    types, 119, 121, 210
    uses, 119–21, 209
Strength of materials, 1, 10, **G311**
Strengths of materials (*see also* Reactions, as material strengths):
    compression per unit area, 220, 221, 223
    load capacity, 125, 126, 141, 156–60,183, 216
    ultimate, **125**
    yield, 124
Structures:
    analysis of, **1**, 10, 11
    integrity of, **1**, 157–60, 213
    members of:
        primary, **153**
        secondary, **153**

Tensile strength (*see* Reactions, as material strengths, tensile)
Tension (*see* Forces, types, tension)
Thermal shock, **120**
Torsion (torque) (*see* Forces, types, torsion)
Torsional Strength (*see* Reactions, as material strengths, torsional)

Ultimate strength (*see* Strengths of materials, ultimate)
Uniformly distributed loads (*see* Loads, distribution of, uniform)

Vectors, **17**
    component form, **18**, 19, **20**, **21**, 22
    direction of, **17**, 20, 23
    magnitude of, **17**, 23
    orientation, 20
    point–force diagrams, **21**
    resultant form, 22, **24**
    sense of, 10
    sums of:
        colinear, 9, 20
        noncolinear, **20**

Wall footing (*see* Concrete, footings, wall)
Wood:
   allowable deflection, 188
   beams, 179–184, 202–3
   columns and posts, **185**, 198
     estimation of sizes, 201
   decking, **184**
     design, 195–97, 199
   flooring, **184**
   joists, **184**
     end, **185**
     header, **185**
     load-bearing capacities of, 190–91, 192–94
     sizing of, 187–190, 192–94
   milling in, 186
   moment of inertia, 179
   nomenclature, **184**, 185
   subflooring, **185**
Work, **39**, 40, 82
   scenarios:
     multiple force/constant distance, 43–45
     multiple force/multiple distance, 45
     single force/constant distance, 39–40
     single force/multiple distance, 40–42, 47
   units of, 39

Yield point, **124**
Yield strength (*see* Strengths of materials, yield)

Zero axis, **181**, 182 (*see* also Centroids, of circles, of triangles)

UNIT CONVERSIONS

Weights and Volumes
1 lb = 16 oz
1 kg = 2.21 lb
1 gal = 8 lb
1 gal = 231 in^3
1 gal = 0.137 ft^3
1 bf = 144 in^3
1 bf = $\frac{1}{12}$ ft^3 (0.0833)
1 kip = 1000 lb
1 ton = 2000 lb

Areas
1 in^2 = 6.452 cm^2
1 ft^2 = 144 in^2
1 ft^2 = 0.093 m^2
1 yd^2 = 9 ft^2
1 yd^2 = 0.836 m^2

Area Formulas
$\frac{1}{2} b \times h$ (triangle)
s^2 (square)
$l \times w$ (rectangle)
$\pi \times r^2$ = (circle)
$2.598 \times s^2$ (hexagon)
$4.828 \times s^2$ (octagon)
b = base
h = height
l = length
r = radius
s = side length
w = width

Decimal Prefixes
kilo-: 10^3 = 1000
hecto-: 10^2 = 100
deka-: 10^1 = 10
deci-: 10^{-1} = 0.1
centi-: 10^{-2} = 0.01
milli-: 10^{-3} = 0.001

Velocity
1 mph = 1.47 fps

Fractions to Decimals
$\frac{1}{16}$ = 0.0625
$\frac{1}{8}$ = 0.1250
$\frac{3}{16}$ = 0.1875
$\frac{1}{4}$ = 0.2500
$\frac{5}{16}$ = 0.3125
$\frac{3}{8}$ = 0.3750
$\frac{7}{16}$ = 0.4375
$\frac{1}{2}$ = 0.5000
$\frac{9}{16}$ = 0.5625
$\frac{5}{8}$ = 0.6250
$\frac{11}{16}$ = 0.6875
$\frac{3}{4}$ = 0.7500
$\frac{13}{16}$ = 0.8125
$\frac{7}{8}$ = 0.8750
$\frac{15}{16}$ = 0.9375

Moments of Inertia
1 ft^4 = 20,736 in^4
1 cm^4 = 0.024 in^4
1 m^4 = 2,402,490 in^4

Work
1 ft-lb = 0.138 kg-m
1 ft-lb = 13,800 g-cm
1 kip-ft = 1000 ft-lb

Distance
1 cm = 0.394 in
1 ft = 12 in = 0.305 m
1 yd = 3 ft = 0.914 m
1 mi = 5280 ft

Unit Force
1 psi = 70.293 g/cm^2
1 psi = 144 psf
1 psi = 702 kg/m^2
1 psf = 4.875 kg/m^2
1 ksi = 1000 psi

Unit Work
1 ft-lb/in^2 = 2142.40 g-cm/cm^2
1 ft-lb/in^2 = 144 ft-lb/ft^2
1 ft-lb/in^2 = 214.24 kg-m/m^2

Miscellaneous
Douglas fir density = 2.65 lb/bf
Steel modulus of elasticity = 29,000,000 psi
Allowable deflection values:
   Steel = $\frac{1}{360}$ = 0.00278
   Wood = $\frac{1}{240}$ = 0.00417